Advances in Intelligent Systems and Computing 205

Editor-in-Chief

Prof. Janusz Kacprzyk
Systems Research Institute
Polish Academy of Sciences
ul. Newelska 6
01-447 Warsaw
Poland
E-mail: kacprzyk@ibspan.waw.pl

For further volumes:
http://www.springer.com/series/11156

Pierre-Jean Benghozi, Daniel Krob,
and Frantz Rowe (Eds.)

Digital Enterprise Design and Management 2013

Proceedings of the First International Conference on Digital Enterprise Design and Management DED&M 2013

Springer

Editors
Pierre-Jean Benghozi
CNRS - Ecole Polytechnique
PREG-CRG
France

Daniel Krob
Ecole Polytechnique
LIX / DIX
France

Frantz Rowe
Institut d'Economie et Management
de Nantes
Université de Nantes
Chemin de la Censive du Tertre
France

ISSN 2194-5357 ISSN 2194-5365 (electronic)
ISBN 978-3-642-37316-9 ISBN 978-3-642-37317-6 (eBook)
DOI 10.1007/978-3-642-37317-6
Springer Heidelberg New York Dordrecht London

Library of Congress Control Number: 2013934119

Printed on acid-free paper

Springer is part of Springer Science+Business Media (www.springer.com)

Preface

Introduction

This volume contains the proceedings of the first International Conference on « Digital Enterprise Design & Management » (DED&M 2013 ; see the conference website for more details: `http://www.dedm2013.dedm.fr/`).

The DED&M 2013 conference was jointly organized by the Dassault Aviation – DCNS – DGA – Thales – Ecole Polytechnique – ENSTA ParisTech – Télécom ParisTech "Engineering of Complex Systems" chair, the Orange – Ecole Polytechnique – Télécom ParisTech "Innovation & Regulation" chair and by the non profit organization C.E.S.A.M.E.S (Center of Excellence on Systems Architecture, Management, Economy and Strategy) from February 12 to 13, 2013 at the Jardins de l'Innovation of Orange in Paris (France).

The conference benefited of the permanent support of many academic organizations such as Ecole Centrale de Paris, Ecole Polytechnique, Ecole Supérieure d'Electricité (Supélec) and Télécom ParisTech which were deeply involved in its organization. A special thanks is also due to Agirc-Arrco, BizzDesign, BNP Paribas, Bouygues Telecom, MEGA International and Orange companies which were the main professional sponsors of the conference. All these institutions helped us to make the DED&M 2013 a great success.

Why a DED&M Conference?

The Digital Enterprise begins to emerge, but real changes that will bring digital business models and digital processes at the heart of organizations are still to come. There is a real stake, on the one hand for the professional organizations that must understand this evolution and appropriate it as a genuine enterprise model and on the other hand, for the academic world to form the actual skills upstream and develop research activities focused on key digital challenges. This is why mastering digital systems requires an integrated understanding of professional practices as well as sophisticated theoretical techniques and tools.

To do so, we believe that it is crucial to create an annual *go-between* forum at international level, opened to all academic researchers and professional practitioners who are interested in the design and the governance of digital systems from an Enterprise Architecture perspective. The "Digital Enterprise Design & Management (DED&M)" conference meets exactly this objective. It aims to become the key place for international debates, meetings and exchanges on the Enterprise

Architecture dimension of the digital business. Our event namely intends to put digital issues at the heart of its program, but also to bring together all business and technological stakeholders of the Digital Enterprise.

The DED&M conference scope integrates consequently both the digital customer & business dimensions (new digital customers behaviors, digital strategies, proposal and distribution of digital value, digital marketing, digital resources management and governance, digital corporate partnerships, etc.) and the underlying technological dimension (information & communication technology, information systems architecture, database & software engineering, systems and networks engineering, etc.).

The DED&M Academic-Professional Integrated Dimension

To make the DED&M conference this convergence point between the academic and professional digital enterprise communities, we based our organization on a principle of parity between the academic and the professional spheres (see the conference organization sections in the next pages). This principle was implemented as follows:

- all Conference Committees (Organizing, Program and Strategic) consisted equally of academic and professional members,
- Invited Speakers are also coming equally from academic and professional environments.

The set of activities of the conference followed the same principles: it is indeed a mix of research seminars and experience sharing and academic articles & professional presentations. The conference topics covers in the same way the most recent trends in the digital enterprise field from a professional and an academic perspective, including the main professional domains and scientific and technical topics.

The DED&M 2013 Edition

The DED&M 2013 first edition received 25 submitted papers out of which the Program Committee selected 12 regular papers to be published. Only the best papers were selected by the Program Committee in order to guarantee the high quality of the presentations. 2 complementary invited papers written by the conference invited speakers were also published in these DED&M 2013 proceedings.

Each submission was assigned to at least two Program Committee members who carefully reviewed the papers (in many cases with the help of external referees). These reviews were discussed by the Program Committee during a physical meeting held in C.E.S.A.M.E.S. in October 2012 and via the Easy Chair Conference management system.

We also choose 12 outstanding speakers with various professional and scientific expertises who gave a series of invited talks. The first day was dedicated to the presentation of **new digital business models**. It consisted of 6 high-profile invited seminars in order to give to the participants a clear, synthetic and large vision of the domain. An open discussion with the invited speakers as well as series of contributed presentations were completing this first day that ends into the conference diner in La Cité de la Céramique de Sèvres. The second and last day of the conference was devoted to new **digital practices and technologies**. Six invited speakers as well as contributed presentations were illustrating this theme. "Best papers awards" were announced at the end of the day by the Scientific Committee chairmen and by the president of C.E.S.A.M.E.S. A farewell cocktail finally ended the conference.

Acknowledgements

We would like finally to thank all the members of the Strategic, Organizing and Program committees for their time, effort and contributions to make DED&M 2013 a top quality conference. A special thank is addressed to C.E.S.A.M.E.S. non profit organization team as well as to SW & Vous which managed permanently with an huge efficiency all the administration, logistics and communication of the DED&M 2013 conference (see `http://www.cesames.net`).

The organizers of the conference are also greatly grateful to the following sponsors and partners without whom the DED&M 2013 would not exist:

Academic Sponsors

- CEISAR
- Ecole Centrale de Paris
- Ecole Polytechnique,
- Ecole Supérieure d'Electricité (Supélec)
- ENSTA ParisTech,
- Institut Carnot Mines
- Télécom ParisTech

Professional Sponsors

- Agirc Arrco
- BizzDesign
- BNP Paribas
- Dassault Aviation
- DCNS
- DGA
- Mega International
- Orange
- Thales

Institution Sponsors

- Digiteo Labs
- Ministère de l'Enseignement Supérieur et de la Recherche

Supporting Partners

- AIM
- Bouygues Telecom
- CNAM
- System@tic
- The Open Group
- Université de Nantes

Paris, February 13, 2013

Pierre-Jean Benghozi – Ecole Polytechnique
Daniel Krob – Ecole Polytechnique
Frantz Rowe – Université de Nantes

Conference Organization

Conference Chairs

- General Chair
  - Daniel Krob, institute professor, Ecole Polytechnique, France
- Organizing Committee Chair
  - Pierre-Jean Benghozi, Research director CNRS, Professor Ecole polytechnique France
- Program Committee Chairs
  - Frantz Rowe, professor, Université de Nantes, France
  - Marc Fiammante, distinguished engineer, IBM Software Group, France

Program Committee

The Program Committee consists of 8 members (academic and professional): all are personalities of high international visibility. Their expertise spectrum covers all the conference topics. Its members are in charge of rating the submissions and selecting the best of them for the conference.

Members

- Co-Chairs
  - Frantz Rowe, professor, Université de Nantes, France (academic co-chair)
  - Marc Fiammante, Distinguished Engineer, IBM Software Group, France (professional co-chair)
- Members
  - Alain Buzzacaro, France télévisions Éditions Numériques, France
  - Brian Cameron, Pennsylvania State University, USA
  - Eric Cohen, Thales Global Services, France
  - Sylvain Lafrance, HEC Montréal, Canada
  - Laurent Leboucher, France Telecom, France
  - Jean-Yves Lignier, AEFE / Club des Maîtres d'ouvrage, France
  - Jerry Luftman, Global Institute for Information Technology Management, USA
  - Stuart Madnick, Sloan School, MIT, USA

- Florian Matthes, TUM, Germany
- Yves Pigneur, Université de Lausanne, Switzerland
- Colette Rolland, Université Paris 1, France
- Dinesh Ujoodah, Société Générale, France

Organizing Committee

The Organizing Committee consists of 8 members (academic and professional) in charge of the logistical organization of the Conference.

- Chair
  - Pierre-Jean Benghozi, Research director CNRS, Professor Ecole polytechnique France
- Members
  - Isabelle Demeure, Télécom Paristech, France
  - Jean-Luc Lucas, ex Orange – France Télécom, France
  - Jean-René Lyon, CEISAR & MphasiS WYDE, France
  - Hervé Panetto, University of Lorraine/CNRS, France
  - Jacques Printz, CNAM, France
  - Pierre-Frédéric Rouberties, CEISAR, France
  - Michalis Vazirgiannis, Athens University, Greece

Strategic Committee

The Strategic Committee helps to define the strategic orientations of the conference. All its members are coming from top executive management of worldwide leading organizations.

- Chair
  - Jean-Christophe Lalanne, Deputy CIO, Air France, France
- Members
  - Jean-Max Arbez, Boost, Switzerland
  - François Bourdoncle, EXALEAD, France
  - Yves Caseau, Bouygues Telecom, France
  - Daniel Dardailler, W3C, France
  - Marko Erman, Thales Research and Technology, France
  - Hervé Gouëzel, BNP Paribas, France
  - Françoise Mercadal-Delasalles, Société Générale, France
  - Grégoire Postel-Vinay, DGSIC, France

Invited Speakers

DAY 1: New Digital Business Models

- Nicolas Colin, Inspecteur des Finances, Inspection Générale des Finances – France
- Dominique Cuppens, Information Systems Director, RFF – France
- Fernando Iafrate, Senior Manager Business Intelligence, EuroDisney – France
- Sylvain Lafrance, Professor, HEC Montreal – Canada
- Jerry Luftman, Professor, Global Institute for Information Technology Management – United States
- Jean-René Lyon, CEO, MphasiS Wyde – France

DAY 2: New Digital Practices & Technologies

- Richard Baskerville, Professor, Georgia State University – USA
- François Bourdoncle, CTO, Exalead - Dassault Systèmes – France
- Dietmar Fauser, Director Software Development, Amadeus – France
- Hervé Panetto, Professor, University of Lorraine / CNRS – France
- Patrick Starck, President – CloudWatt – France
- Paul Timmers, Director Sustainable & Secure Society, European Commission – Belgium

Contents

Use Case: Business Intelligence "New Generation" for a "Zero Latency" Organization (When Decisional and Operational BI Are Fully Embedded)

Fernando Iafrate

1 Introduction

Business Intelligence link to an EDA (Event Driven Architecture) for a "Zero Latency Organization"

2 What for?

"To Serve Every Day Thousand of Dreams"
When the operational performance is a key success factor to deliver the expected customer on site experience (where the dreams come through), the monitoring of this performance in order to anticipate and take the right action is mandatory.

The Business Intelligence link to and Event Driven Architecture in conjunction with business process, is the corner stone of the monitoring of the on site activity.

This is achieved via predictive analysis, near-real-time traffic monitoring and performance management.

3 Key Issue

What are the factors that lead to successful strategic deployment of business intelligence and information management?

Fernando Iafrate
Senior Manager Business Intelligence,
EuroDisney
France

P.-J. Benghozi, D. Krob, & F. Rowe (Eds.): *Digital Enterp. Des. & Manage. 2013*, AISC 205, pp. 1–7.
DOI: 10.1007/978-3-642-37317-6_1

4 What You Need to Know

Disneyland Paris built a world-class Business Intelligence environment, which provides customer-focused strategic and near-real-time operational insight to a broad set of users. Business activity is predicted and continuously monitored against key performance indicators (KPI's). Unexpected traffic patterns and congestion in the parks are quickly identified and addressed. As with most successful BI projects, best practices were involved in the project, such as a strong focus on business problems over technology, sticking to pre-defined infrastructure standards and reliance on a BI competency center.

Disneyland Paris, located near Paris, was opened in 1992 and has become the most frequented tourist destination in Europe, with over 16 million (in 2012) visitors per year. Its business goes beyond the theme parks to include hotels, shops and restaurants.

Disneyland Paris has long understood that customer satisfaction is a key business metric.

Each day is different at Disneyland Paris. While crowds move in predictable patterns that vary with the season and day of the week, a cloudy day or sudden rain storm can create unpredictable traffic flows.

Business Technology (IT) department at Disneyland Paris embarked on an OPM (Operational Performance Management) project to provide "real time" information and alerts about operational business activity to the operational managers, so they could become aware of urgent conditions, make fact-based decisions and take immediate action. The project was a partnership between the BT organization and line-of-business users. An OCC (Operation Control Center) and distribution system was planned to monitor the global operational activity. To Disneyland Paris, building the OPM application was more than just a project — it was a cultural change that embraced real-time information as a way for it to become more customer-centric and to be better achieving its motto **"to serve every day, thousands of dreams**."

5 Challenge

Disneyland Paris was faced with a challenge was driven by technical and business issues, but also a change in operational activity monitoring culture (move from an on site manager skill/local context based decision to a solution fact based decision monitored by the OCC).

- A key business requirement was to reduce customer waiting time wherever possible.
- Cultural changes were driven by a workplace that required action based on, and workers were made more accountable for their **zero-latency decision** (is a decision where the action latency is align on business process timing) operational efficiency, which was continuously measured. Along with increased awareness came a degree of autonomous authority to correct problems at the local level. In effect, everyone was to be made responsible for customer satisfaction.

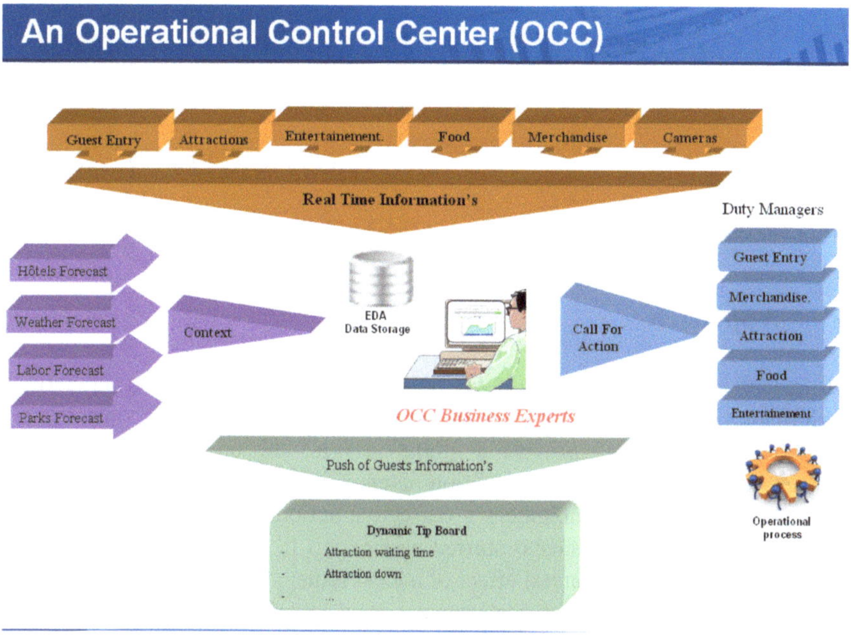

Fig. 1 The Operational Control Center

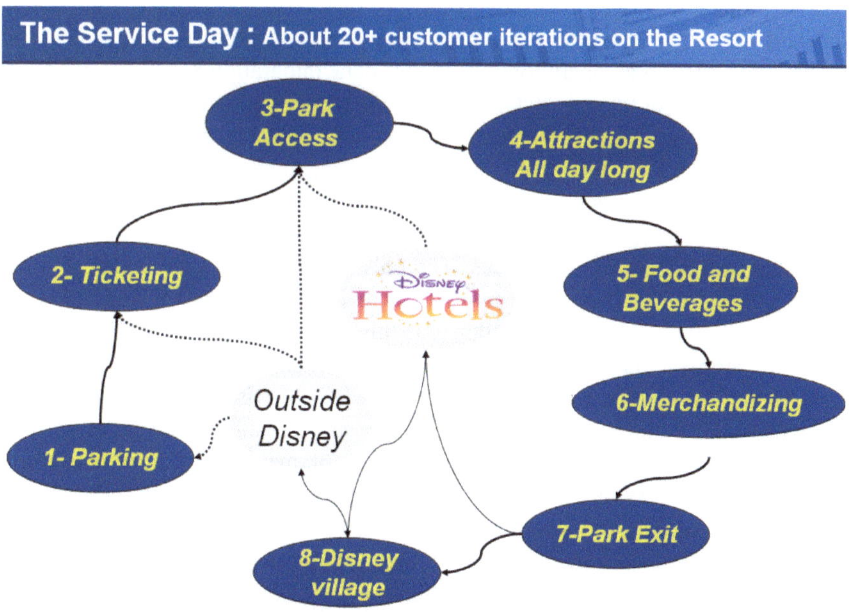

Fig. 2 On site customer interactions

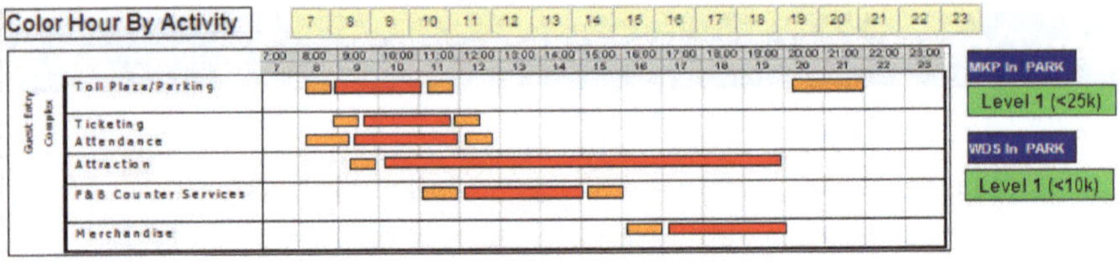

Fig. 3 The daily activity workflow by peak period (in orange & red)

6 Approach

Disneyland Paris took a multilevel approach to implement the operational performance management solution and processes. An operational control center was created in order to provide a global view of the on site activity (parks, restaurants, shops…). The Information latency is aligned to business process timing (time needed to implement a corrective action), which is near real time. The operational control center is staffed with expert from each line of business in order to anticipate the collateral effect of a local issue.

As part as the operational performance management solution, the employees can have access to their operational performance metrics against the goals of their own department and others. The solution is also available on mobile devices, allowing the managers to receive alerts and high-level reporting.

The project leverages the same BI tools that Disneyland Paris uses for its standard reporting solution. An engineering study confirmed that the system components (network, operational systems, data integration tools and reporting tools) would perform well under low-latency conditions (moving less data but more frequently). Data is gathered from hotels, ticket windows, food service outlets, merchandise stores and attractions. Acceptable report performance is achieved by accessing some of the data directly from transaction systems, and other from the decisional systems. The data model for operational reporting has been standardized across all sources. The only raw data collected is date & time, location, transaction count and associated revenue. All metrics and KPIs are derived from the simple data model, but this leads to powerful indicators. For example, shop productivity is determined by correlating the percentage of cash drawers that are opened within a five-minute window, with the number of people that entered the shop. Short-term history is maintained by the solution. The business rule logic performs threshold analysis of metrics and generates alerts. Overnight, the history from that day is loaded into a data mart for use in predicting future traffic patterns and a new local modelization & forecast is executed every hour, in order to realign the targets with the current operational environment (i.e. variance in term of parks attendance)

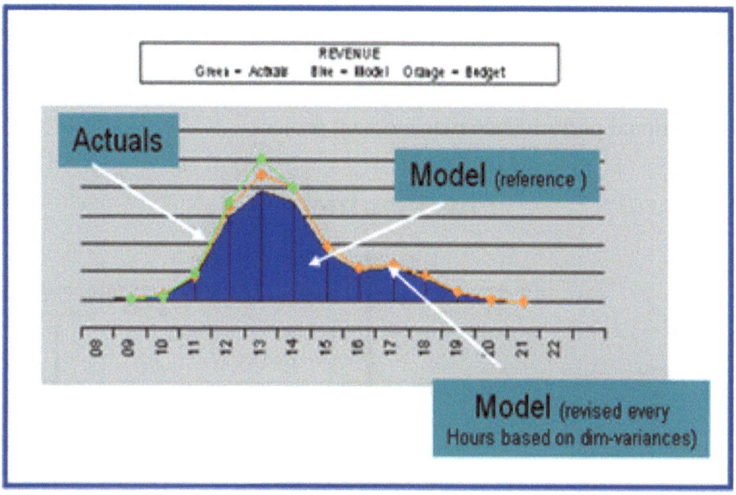

Fig. 4 Dash board showing the current activity (actuals) against the predictive models

Predictive modeling is an important aspect of the performance monitoring system. Traffic history and modeling tools are used to predict the crowd volume and flow for each hour of the day. Incoming hotel reservations are part of the model, as this provides a projection several weeks into the future. The model is updated each night, and again at each hour while the park is opens. With hourly updates, the models become more accurate as the day progresses and the impact of unusual patterns can quickly be understood, and adjustments made.

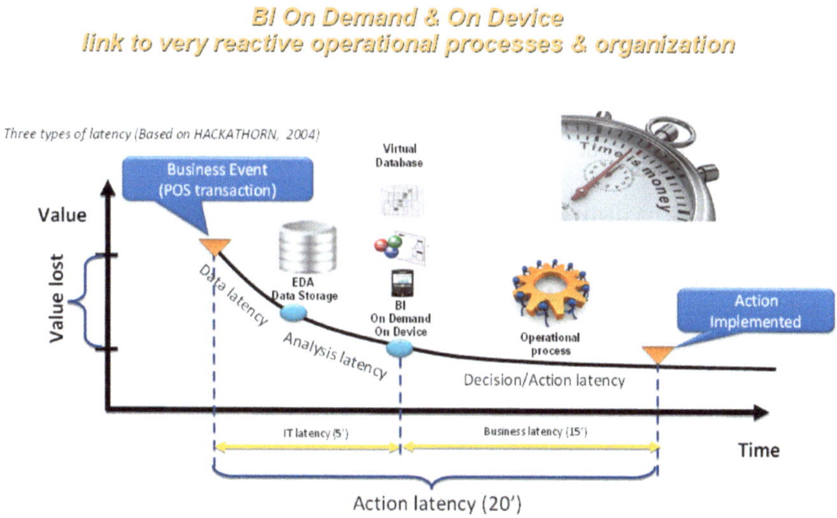

Fig. 5 BI a question of timing & latency

7 Benefits

Customer satisfaction is the primary performance indicator that measures success of the operational performance management project.

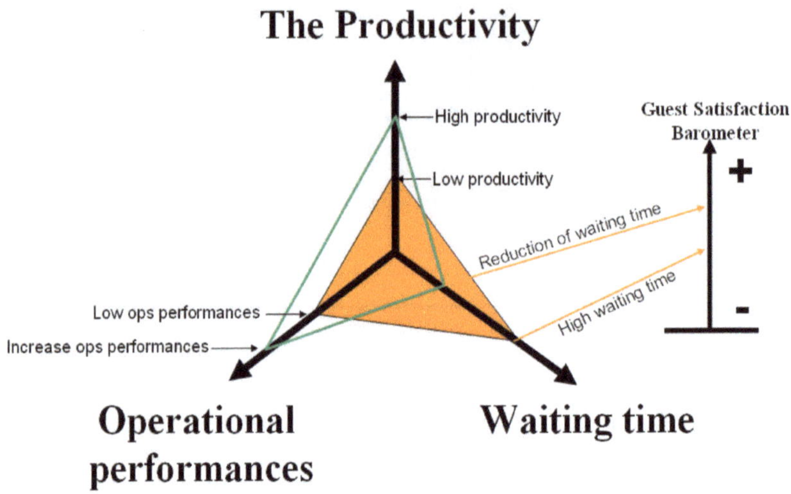

Fig. 6 Increase of the guest satisfaction

The democratization of performance information has changed the culture of Disneyland Paris. The operational performance can be measure against a predicted goal at all times.

8 Critical Success Factors

- Create an information democracy, where information is used and acted on across the organization.
- Provide trusted predictive modeling with sufficient lead times to plan resource schedules.
- Reuse existing IT investments wherever possible while providing sufficient scale and flexibility.
- Define and measure KPIs that align with strategic goals.

9 Lessons Learned

- Predictive modeling and operational information are both required. Modeling helps to anticipate a situation and put operational information into context.

- Process changes are required to gather data in near real time and make it available to decision makers.
- Patience is required to see the potential value, but it is also important to get something useful running quickly, to get the culture changes started and provide continuous value.
- It is important to focus on the process more than the technology. While the reports show baselines, predictions and actual activity, the real value comes from adjusting the business processes and using BI tools to analyze the impact of efficiency experiments.

The Enterprise as the Experiential Design Platform

Richard Baskerville

Abstract. Individuals are autonomously designing, creating, and operating their own complex, idiosyncratic information systems. These systems are designed in an experiential and emergent way. This growing individual technological autonomy creates dynamic bindpoints where formerly there were relatively static user endpoint interfaces across an air gap. This enables the organization to adjust the location of its enterprise system bindpoints in relation to the individual system endpoints.

1 Introduction

In the past there were few who would question that the rightful ownership of an organization's information system would be organization itself. However the increasing availability of information technologies to the individual is providing powerful individually designed information systems that are increasingly available to interact with organizational information systems. When workers incorporate their own complex information systems into their daily work processes, their organizations begin to shift from a platform that was entirely owned by the organization to a platform that is partly owned by the organization and partly owned by the individual. Processing on such shared systems is not a new idea for many organizations today. Shared computing platforms across supply chains have meant that an organization's information system is sometimes shared with other organizations. The reliance on cloud computing and software-as-a-service has also increased the sharing of computing between organizations (Fingar, 2009).

Richard Baskerville
Department of Computer Information Systems,
Georgia State University,
POB 4015,
Atlanta, Ga, 30302 USA
e-mail: baskerville@acm.org

P.-J. Benghozi, D. Krob, & F. Rowe (Eds.): *Digital Enterp. Des. & Manage. 2013*, AISC 205, pp. 9–21.
DOI: 10.1007/978-3-642-37317-6_2

The arrival of a worker supplied information resources is also not new. Although there have been nagging security problems, workers have frequently brought their own devices for use in their organizational work life (Kennedy, 2013).

Increasingly these worker supplied information resources are growing into complex multi-device, multi-software architectures in their own right. The result is a set of information resources shared between the organization and its workers. Importantly there is a potential collision between the architecture of the organization's information system and the architecture of the individual worker.

The arrival of complex individual information systems is new. Of particular novelty is the manner in which these systems are designed and developed. The architectures of these systems are not developed by trained system professionals for a persistent, effective and reliable information platform. Rather, they emerge through a process of design-experience-redesign that places the organization into a dynamic architectural setting.

In this paper we explore the interaction between individual information systems and organizational/enterprise information systems. This interaction is increasingly important because it is increasingly difficult to prevent workers from creating the interaction between the organization's systems and their own individual systems (Luftman et al., 2012). This interaction is also important because it offers organizations powerful new information resources developed and made available as part of their employment of smart, technologically adept workers.

2 Individual Information Systems

Individual Information Systems (IIS) are, in the first place, information systems. We distinguish these from the more common notion of organizational/enterprise information systems (OEIS). In this paper we will take the view that information systems are complex social technical phenomena (Bostrom & Heinen, 1977; Mumford & Weir, 1979). These systems arise in a human computer system. It is information and communications technology in a human context. Information systems are not just an accumulation of technology, information, and human factors. Information systems are a synergetic whole that results in a symbiosis of information and communication technology with information and human factors that produce more than just the sum of the parts.

Stephen Alter (2008) inventories more than 20 important definitions of information systems. While many are similar, this suggests that there are many differences in the way experts define information systems. Of the 20 definitions, most include references to computer or technology. Most of these definitions also refer to organizations in some way. Some of these definitions mention society or social aspects, and most would exclude an individually owned information system. The proposal in this paper, that individuals have now become so empowered by technology that they are operating an information system as an individual rather than as an organization.

Since we are obliged to have a working definition of information systems for our purpose in this paper, we can adapt Alter's (2008) definition for our purposes: an information system is a type of system, "in which human participants and/or machines perform work (processes and activities) using information, technology, and other resources to produce informational products and/or services for internal or external customers".

We have specifically adopted the term *IIS* rather than *personal information systems* because the term *personal information system* is already present in the literature with a much narrower scope of meaning. Information science has defined personal information systems from a database or bibliographic perspective (Burton, 1981, p. 440) for the support of personal "collections and personal indexes" (Moon, 1988, p. 265). Personal information systems are defined as,

A "personal information system [is one that] provides information tailored to an individual and delivered directly to that individual via a portable, personal information device (PID) such as a personal digital assistant, handheld PC, or a laptop." (Silberschatz & Zdonik, 1996, p. 770)

Personal information systems are regarded as individuals with a particular technology being serviced in their needs for data or information. The prevailing definition does not allow recognition of complex multi-technology information systems under the control of an individual.

However there is one important aspect of the work in personal information systems that is relevant to any definition of IIS. This is the idiosyncratic nature of personal information systems that also will inhabit IIS (Burton, 1981). This is because IIS, like personal information systems, correspond to unique individuals.

By integrating these two somewhat contradictory streams, namely, information systems, and personal information systems, we can derive a working definition of an IIS:

> An IIS is a system in which individual persons, according to idiosyncratic needs and preferences, perform processes and activities using information, technology, and other resources to produce informational products and/or services for themselves or others.

3 Example: Sam Spade

Many readers will have experience with their own IIS, and may find it surprising to argue that their own personal computer (PC) or laptop should be elevated to the lofty conceptual level of an IS. Before discussing how this form of IS has evolved, we should examine a case of this phenomenon (reported in 2011a; Baskerville, 2011b). Keep in mind that IIS follow idiosyncratic needs and preferences, so no immediate case can be considered typical.

Sam Spade (a pseudonym) is a professional employee in a large government division. Spade has three computers to his personal use: two desktop machines and one laptop. His employer has located one desktop in Spade's office on the employer's premises. Spade owns the other desktop, which is in his home office. Spade's employer also provides a laptop for his personal use. Spade also owns two other laptops he shares with his family. At home, Spade also owns and uses a smart phone, and a combination printer, scanner, and fax machine. All of the devices in his home are networked into a local area network (LAN) that includes a DSL modem, a firewall, an Ethernet router, and a wireless access point. He uses three Internet providers: the DSL connection in his home via a telephone provider, an Internet link to the smart phone via his mobile phone provider, and the connection to his office using his employer's LAN.

Spade has installed more than 50 separate software packages on these computers. His main activities involve only a few of these. The mainstay of his work life is the productivity software package with its spreadsheet, presentations, and especially the word processing tool. He uses this package to generate documents in all facets of his profession. In connection with this tool, he uses accessory writing packages like dictionaries. He also depends on an email package as his main communications medium, and uses a diary/calendar application for planning and record-keeping. He uses a VOIP package for low cost teleconferencing across the Internet. The data related to these major packages are synchronized between PCs, laptops, and his smart phone.

Spade accesses services related to his profession from a cloud provided by his employer. The term cloud is used here in its loose, IS perspective because, in terms of access to services, the cloud is evolving. This notion is broader than cloud computing and extends to cloud-based processes and information systems. This evolution represents the increasing availability of cloud-based business processes as well as low-level data services (Fingar, 2009).

This cloud permits access to many reference resources, such as publications and regulations, much of which is contracted by his employer. The employer-provided cloud also provides online access to customer and vendor data, online professional tools, and virtual meeting resources. The employer's personnel unit allows Spade online access his personnel and compensation records. Spade also uses a few additional professional services from outside his employer's cloud, including writing aids, discussion groups, meeting planning, and shared file folder drop sites.

In terms of his personal business, a growing cloud of personal finance services is supported by the retail banking, insurance, and financial services industry for individual customers. Spade uses a personal finance package to harvest services from this cloud, and to manage banking accounts and credit cards. He downloads and synchronizes transactions from his accounts for reconciliation with his records. He uses a portfolio package to manage a shares/stock investment portfolio that spans several brokerage and insurance investment accounts. In addition, he uses income tax software to prepare annual tax reports. The tax and personal finance software synchronizes automatically, drawing information from the various clouds. Results are filed directly with the tax authorities across the Internet.

Spade's personal and professional lives overlap where his employee financial transactions touch his personal accounts, as in the case of compensation and expense reimbursement. The overlap extends to travel expenses because Spade does most of his travel planning through airline and hotel booking web sites via the Internet. He often shops online, and while most of his online purchases are personal, he sometimes purchases business related items and claims reimbursement. Such professional expenses go through Spade's personal books further extending the overlap between his employer's IS and his own individual IS.

Spade makes the parts of his IIS that he owns available to his family. Family members share the home LAN for email communications and social networking services. The family also has access to Internet-based films, videos, and television programs that form part of the subscription service from the local cable television provider.

Figure 1 details this IIS architecture. The architecture in this case demonstrates how this individual's usage of ICT has evolved beyond the boundaries of personal computing or ICT alone. These are simply the bottom two layers that provide the foundation for the information architecture. Here we find work and activity flows, business and personal information processes, and social technical design decisions. These elements form IS problems that are similar to those of large businesses or SMEs. The PC has grown into an IIS that is a genuine IS. Because such IIS architectures are idiosyncratic, other examples would be more complex, and yet others simpler.

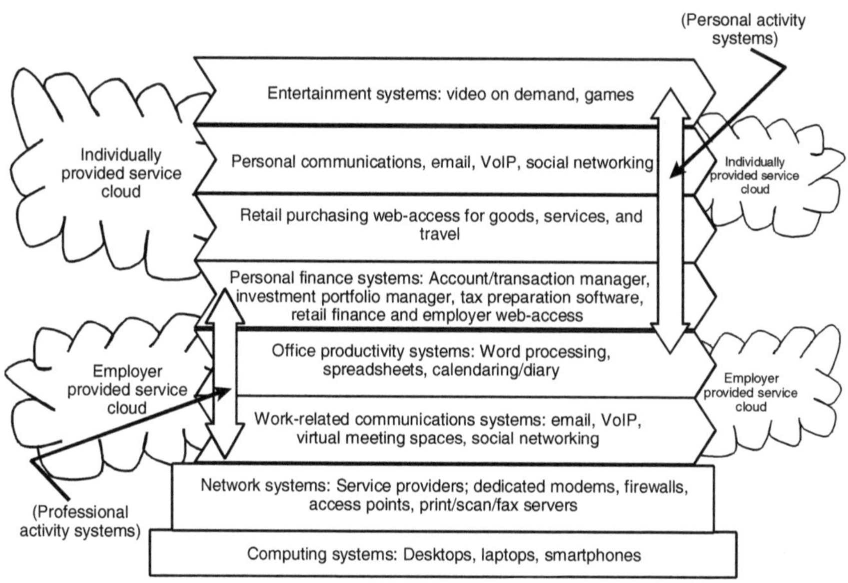

Fig. 1 Spade's IIS architecture (adapted from Baskerville, 2011b)

There are two vertical arrows that denote two overlapping activity systems that form subsystems within this IS architecture. The professional activity system corresponds to Spade's information processing activities in his role as an employee. The personal activity system spans Spade's information processing activities outside of his role as an employee. The arrows cut across architectural elements denoting how the distinction between the elements can blur, such as in cases where the employer provides individuals with personal Internet access. The figure also represents how Spade is consuming information services and producing information that arise from, and sinks into, two clouds that do not necessarily overlap. One cloud is generally provided by his employer for use in his professional activities. The other cloud is constructed of services (such as retail financial services) for which Spade contracts individually.

4 Individual Technological Autonomy

Technological autonomy is more often considered as a characteristic of a nation state or culture (McOmber, 1999). For our purposes, we are more interested in technological autonomy as a characteristic of an individual refers to the degree of independence with which workers within an organization command the architecture of their own IIS. Many organizations are contending with the presence of workers' own devices in the workplace; often referred to as *bring your own device* (Kennedy, 2013). This viewpoint suggests that the OEIS must now be configured in order to interact with a particular device provided by the individual workers. It is a simplistic notion that the worker will simply provide the organization with their preferred device. It fails to recognize that such devices might better be acknowledged as portals to the individual's information system. In other words, such a device is an interface between the OEIS and the IIS. This may be a more realistic representation of the situation as *bring your own system.*

A worker-provided interface to their own IIS is rather a more complex proposition than merely providing the organization with a device. The organization is acquiring an interface to a system with a great degree of individuality and autonomy. We cannot assume that people are simply organizational system nodes. People have idiosyncratic reasons for using the OEIS. These reasons involve idiosyncratic adoption patterns, such as which feature subset will be operationally adopted. Idiosyncratic adoption patterns have created issues in the study of adoption because it is difficult to measure adoption in settings where each individual adopts a system differently. As a result a great deal of information systems research into adoption uses measures of intention to adopt rather than adoption itself. Such measures are chosen because the intention to adopt is a simpler and more easily measured construct. There are also issues of individuality in end user development where each individual does more or less in the development of the system. And there are also issues with task technology fit that substitutes a suite of task performances for the individual.

The research into personal information systems brought recognition that when people use their own systems to interact with libraries they begin storing and processing library and database information in idiosyncratic ways. Their usage of library systems was clearly subjective in both its scope and in its methods or processes for approaching the information. In order to accommodate this idiosyncrasy, OEISs encountered new overhead, for example, accounting for complexity in query and database schemas. This complexity has also been noticed in research into executive information systems that extend their reach into highly diverse information sources for strategy setting.

5 Experiential Design

While IIS may offer the same profile as a complex architected OEIS, these are rarely designed with the same kind of rational design process. IIS have a highly variable scope, idiosyncratic processes, and often weird configurations of information technology. It might be tempting to attribute unusual designs to an uneducated designer and user. However, this design process operates in a different context than the design process of a professionally architected OEIS. IIS have unlimited system availability to their user. As we have seen from the research into personal information systems, IIS have richer information attributes than many OEISs. The design goals for IIS are also different. These design goals often include not only utilitarian goals, but hedonistic goals and desires for particular social outcomes. The design goals may also include value, enjoyment, and even status enhancement. The latter can occur when an IIS is outfitted with a particular technology in order to reflect the desired social status. Research has already shown that Internet usage patterns often reflect high social and community purposes.

IIS are also bounded both socially and geographically to particular sets of information and communication technology. The individual designer is operating within a confined sociotechnical context. Purchase decisions are highly influenced by the individual designer's social and professional networks. Family, neighbors, and coworkers all influence component selection through their own past choices and their advice to the designer about their future choices. The design components are chosen from mass-market devices usually at very low cost. These devices are consequently highly standardized and reconfigured for the most general individual use. Yet the designer is aiming to achieve a very peculiar system by using these standard parts. This puts the architecture into a tension setting. There is a tension between the idiosyncrasy of the designer and the design process and the standardized, preconfigured components available from which to construct the IIS. As a result such systems often have complicated and inefficient workarounds to make the various components work together to achieve the rather peculiar goals and aims of the designer.

This context creates a setting for a design process that involves an explorative patchwork approach to designing an IIS architecture. It is a situation in which the designer is both designing and experiencing the design results at the same time (Baskerville, 2011a). This process of experiential design is important in accommodating interactions with organizational systems because of its idiosyncratic results and continuous adaptation. Experiential design is necessarily explorative because system components are continuously being disconnected and reconnected in different ways. It is a trial and error process that sometimes involves acquiring temporary components such as trial software or returnable devices in order to experiment with different designs. The design process interacts with the artifact being designed. There is a hodgepodge of components that are progressively acquired for integration into the IIS. This hodgepodge is worsened because these components are acquired piecemeal and often without a long-term plan. Because components cannot be easily discarded and replaced, the component hodgepodge places a more severe constraint on the redesign of the IIS. This makes for its matching more difficult because the existing hodgepodge system provides the basis for determining new components. This determination arises with any goal elaboration and drives a search for a feasible elaboration of the existing system in order to achieve the goal elaboration. These are highly emergent systems often with very short term planning and operating horizons.

6 Enterprise Individual Bindpoints

Enterprise individual bindpoints are distinct from OEIS end points. They represent dynamic intersections between different information systems. By their nature they shift the system air gap deeper into the IIS. In this section we will deal with each of these concepts: enterprise end points, dynamic intersections, system bindpoints, and air gaps.

An *air gap* is an information security concept that refers to the lack of an electronic connection between two elements of a system (Lindqvist & Jonsson, 1998). Any information to be conveyed from one element to the other element requires some physical intervention, such as keying information into a system terminal. For many OEISs, the air gap occurs at the end user interface. At this point the user interface often involves a video screen, a keyboard, and a mouse. Information is sent to the user on a video screen, and retrieved from the user via the keyboard. Of course the user may also introduce pluggable devices such as a flash memory stick where allowed. In security terms, there is an air gap between the user and the OEIS resources.

A user air gap often represents one end point of an OEIS. An *endpoint* is the mark of a terminus or completion that generates or terminates an information stream. It can involve remote devices such as laptops or other wireless and mobile devices that connect to an enterprise system. Many current enterprise architectures assume that a user endpoint will involve a device and an air gap, such as a laptop connected to a server. It is a fundamental assumption underlying bring-your-own-device (BYOD) thinking.

This notion of an air gapped end point is quite different from a *bindpoint* that represents the new (changed) context created when a new or altered IIS connects with an OEIS. The bindpoint occurs when the enterprise system connects to an IIS. Distinctive to the notion of a bindpoint is the dynamic and changing character of an IIS. As we learned above, an IIS is emergent and idiosyncratic. But the bindpoint does not simply represent the connection between two systems. Owing to the nature of experiential design, the connection will change the architecture of the IIS, and affect the emergent goals, affordances, and constraints that inhabit the system. Upon connection, the designer will redesign the IIS in an experiential way. The IIS becomes a changed and reconfigured system. Because of the connection, the IIS is moved to a new place (a new context) and becomes a new platform for experiential design. Because the connection is a system to system connection, we cannot assume that the OEIS is unaffected by this dynamic connection. Replacing an endpoint with a bindpoint elaborates the enterprise system in a way that users are experientially designing. The enterprise system also moves to a new place (a new context). The result is similar to the notion of a bindpoint in a computer game. When a character in a game encounters or uses an object, both the character and the object may be taken to a new place in the game. The new place is the bindpoint (Griffiths et al., 2003). Importantly bindpoints may be too innumerable to predict every permutation of the elements, but they must be computable.

The connection of an OEIS to an IIS creates a form of cross road or intersection between multiple systems. This intersection is no longer a simple connection of an endpoint that provides a terminus for the enterprise system. It is not merely the entry or exit point for service or process. As a bindpoint it becomes the condition of an IIS after binding to the enterprise system. Concomitantly, it also becomes the condition of the OEIS after binding to the IIS. In settings where many IIS are bound to an OEIS, bindpoints are frequent and diverse. This can mean that the OEIS is subject to constant redesign in the presence of the constant experiential design arising in the constellation of associated bindpoints.

The bindpoint is a new context, not a simple connection. It can affect a diverse array of inputs and outputs for both the OEIS and the IIS. Because the bindpoint changes the context of both the individual system and the enterprise system, it may develop a web of connections between the two architectures. The exact shape of this web is ephemeral to the degree engaged by the dynamics of experiential design (see Figure 2).

Unlike an end point, a bindpoint will usually shift the air gap further away from the OEIS and deeper into the IIS. It can be an important feature of the new context for the enterprise system. The enterprise system's interactions with its user are mediated by experientially designed elements. As the air gap becomes more remote, more governance must be exercised to take best advantage of the more sophisticated input and output processing afforded by the IIS. The functional advantage stands at tension with the security disadvantage.

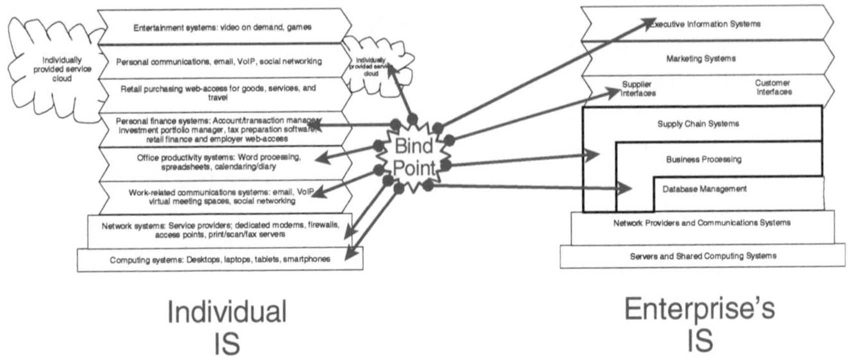

Fig. 2 Bindpoint as a web between individual and enterprise information systems

7 Example: Virtualizing a Digital Forensics Laboratory

In this simple example, we consider the laboratory work of information systems students who are studying digital forensics in a university executive program. The traditional digital laboratory set up would provide one desktop computer for each student in the room and an instructor's desktop computer with a projection screen. A network would connect the computers. Each computer would have the licensed forensics software installed. Students would attend classes in this laboratory, and would perform their homework exercises in the same lab. The student workstations would represent endpoints in this information system architecture, and the air gap would be present at each workstation. In this sense the air gap is always present in the laboratory.

In the executive program all students are provided with a tablet computer and required to provide their own laptop computer to support their studies. All students in the program were self-equipped with smartphones. In this setting the traditional digital laboratory largely duplicated the IIS provided by the students. Further the traditional laboratory constrained the students to complete their work while physically in the lab. An alternative organizational architecture created a lab using virtual machines available for access via remote desktop from student laptop computers. The virtual machines provided the licensed forensics software. For some instances forensics tools had to run in native mode on student laptop computers for purposes such as media imaging that required direct hardware access. Instruction manuals and lab protocols were delivered to student tablets.

The resulting bindpoint created a new context for both the OEIS and the IIS. The organization had to provide ideal wireless connectivity for classrooms in which laboratory exercises would be conducted during class. This was needed to support the remote desktop bandwidth and large file downloading bandwidth for many laptops. The organization also provided a substantial virtual machine server to support the necessary virtual machines simultaneously active during class. Students had to configure or reconfigure their systems to support the course. This meant installing (where required) and configuring the remote desktop software, along with other forensic tools (such as imaging and hashing tools) as required. These tools had to be determined by the students according to the configuration of their own laptop computers and information systems. Because these student systems were idiosyncratic, new demands were placed on the university staff to consult on a wide variety of possible forensic computer configurations. The laboratory protocols had to be substantially revised to provide useful direction for individuals using substantially idiosyncratic system configurations. The air gap moved away from university computers and disappeared somewhere into the IIS.

While this is a simple example, it provides insight into the changing situation when an endpoint is replaced with a bindpoint. In the original laboratory, the university owned and controlled the information system all the way to the air gap. In the revised executive laboratory, the university did not own or control the complete information system. It was partly owned and partly controlled by the university, and partly owned and partly controlled by the student. In the original laboratory, the architecture was persistent to a great extent. The laboratory design was intended to last three years. This meant that there was three years of exploitation from its design. In the revised executive laboratory, the design is intended to last one term (at best). The design and configuration of student IIS evolved in some cases during the course. With each successive term, some students will arrive with the most recently available devices, while others will arrive with devices older than those seen previously. Each idiosyncratic student IIS dictates a new bindpoint for the overall information system shared between students and university. Further, reconfiguration of virtual machines is more easily done from term to term. Rather than lasting three years, the virtual machines are themselves less persistent. It is not only the student IIS that is changing the bindpoints, it is also the OEIS undergoing more frequent change.

8 Discussion

For more than a generation, a steady increase of the proportion of contingent workers in the labor force. This increase is driven by the need for flexibility and the demand for cost competitiveness, but may also arise from the technological independence of workers. Workers, in possession of their own information systems may no longer be dependent on organizations to provide the basic technical resources needed to accomplish their tasks. There are indications that the growth in new positions for contingent workers is outpacing the growth in new positions for standard employees. In 2005 there were more than 2.5 million temporary and contingent workers employed by 90% of large U.S. companies drawing from a market of more than 10 million Americans (Yang, 2012).

Since organizations often regard these trends from a perspective of the benefits to the organization, it is easy to overlook the shifting job security to the worker. Rather than drawing security from a single employer, contingent workers draw their job security from multiple employers. Working with a portfolio of employers can lower the career risk of a contingent worker in comparison to a worker fully employed by a single organization. It is a sense of not having the entire employment benefits in a single basket. Replacing the job-for-life with a career portfolio also satisfies the current shift in worker attitudes toward an emphasis on choice, autonomy, and flexibility (Jorgensen & Taylor, 2008). The ability for knowledge workers to develop multiple employer relationships depends on the IIS because of the necessary technology. However, it also means the worker is engaging multiple bindpoints in their IIS. The web of bindpoints between organizations and knowledge workers is an increasingly complex many-to-many relationship.

Insights into the nature of experiential design in the IIS can also reveal its features in the design of OEIS. Prototyping and agile development adopt experiential design principles to a degree, but in many cases formal resource commitments limit the extent to which experiential design unfolds. Experiential design probably has a higher degree of presence in more independent funded *skunkworks* or unfunded *hobbyshopping* developments, where innovation runs in a management-hands-off setting. End user development within OEIS can also unfold in an experiential design manner. Security event response and disaster recovery often invokes experiential design in early stages, owing to the intense need to restore essential information services to the organization.

The search must begin for practical solutions to the management of this increasing web of highly emergent OEIS bindpoints. Experiential design is probably foremost opportunity for quickly tuning OEIS bindpoints to a moving constellation of IIS. Using this design approach, OEIS developers will have to find ways to manipulate architecturally safe portals into the OEIS. Such bindpoint affordances will themselves have to be marshaled carefully to keep them attuned to diverse IIS bindpoints, and still protect the integrity and security of the OEIS.

9 Conclusion

An IIS is a system in which individual persons, according to idiosyncratic needs and preferences, perform processes and activities using information, technology, and other resources to produce informational products and/or services for themselves or others. These systems are complex, idiosyncratic information systems. Individuals autonomously design, create, and operate these information systems. A number of factors are increasing the organization's need to interact productively with these systems. Factors include the changing technological sophistication of the knowledge worker, and changing trends in worker career and employment options.

Interaction with an array of individual information systems poses a complex problem for organizations with complex information architectures of their own. These individual systems are designed in an experiential and emergent way. They are autonomous and fast-changing. These idiosyncratic features create dynamic system-to-system bindpoints where formerly there were relatively static user endpoint interfaces across an air gap. This change recasts the organizational information endpoint problem. The organization now needs to adjust the location of its enterprise system bindpoints in relation to the endpoints that have moved deep within the individual information system.

References

Alter, S.: Defining information systems as work systems: implications for the IS field. European Journal of Information Systems 17(5), 448–469 (2008)

Baskerville, R.: Design Theorizing Individual Information Systems. Paper presented at the Fifteenth Pacific-Asia Conference on Information System (PACIS 2011), Brisbane, Australia (2011a)

Baskerville, R.: Individual information systems as a research arena (Editorial). European Journal Information Systems 20(3), 251–254 (2011b)

Bostrom, R., Heinen, S.: MIS problems and failures: A socio-technical perspective, part I: The causes. MIS Quarterly 1(1), 17–32 (1977)

Burton, H.D.: FAMULUS Revisited: Ten Years of Personal Information Systems. Journal of the American Society for Information Science 32(6), 440 (1981)

Fingar, P.: Dot Cloud: The 21st Century Business Platform Built on Cloud Computing. MK Press, Tampa (2009)

Griffiths, M.D., Davies, M.N.O., Chappell, D.: Breaking the stereotype: the case of online gaming. Cyberpsychology & Behavior: The Impact of the Internet, Multimedia and Virtual Reality on Behavior and Society 6(1), 81–91 (2003)

Jorgensen, B., Taylor, P.: Employees, employers and the institutions of work: the global competition for terrain in the ageing workforce agenda. Foresight: the Journal of Futures Studies, Strategic Thinking and Policy 10(1), 22–36 (2008)

Kennedy, D.: The Battle of BYOD. ABA Journal 99(1), 26 (2013)

Lindqvist, U., Jonsson, E.: A map of security risks associated with using COTS. Computer 31(6), 60–66 (1998)

Luftman, J., Zadeh, H.S., Derksen, B., Santana, M., Rigoni, E.H., Huang, Z.: Key information technology and management issues 2011-2012: an international study. Journal of Information Technology 27(3), 198–212 (2012)

McOmber, J.B.: Technological autonomy and three definitions of technology. Journal of Communication 49(3), 137–153 (1999)

Moon, C.: Computerized personal information systems for research scientists. International Journal of Information Management 8(4), 265–273 (1988)

Mumford, E., Weir, M.: Computer Systems Work Design: The ETHICS Method. Associated Business Press, London (1979)

Silberschatz, A., Zdonik, S.: Strategic directions in database systems–breaking out of the box. ACM Computing Surveys 28(4), 764 (1996)

Yang, X.: Contingent Worker, Permanent Loser?–How Perceived Trust Shapes Communication between Contingent Workers and Standard Workers in Knowledge-based Organizations. International Journal of Business and Social Science 3(8), 172–180 (2012) and/or interfaces across the gap. This change reflects the organizational [illegible] traditional component problem. The [illegible] to adjust the boundaries of the [illegible]

[illegible]

From a Strategic View to an Engineering View in a Digital Enterprise
The Case of a Multi-country Telco

Hervé Pacault

Abstract. In this paper, we will present two examples of how Enterprise Architecture could help make better investment decisions within a multi-country Telecommunications Company (Telco). These two examples relate to two of the significant challenges that a Telco is currently facing:

- The invasion of Telco's traditional playground by new actors, the web players and the consumer electronics manufacturers : Telcos must react and position themselves face to the new entrants
- The sharing of IT components between the local companies of the same Telco Group, in order to cut cost through economies of scale

For each of the examples, we will present a strategic view and an engineering view, showing that it is possible to show, on a single A4 sheet, a summary of the technical policy for the delivery of end-user services regarding the two above-mentioned challenges.

Keywords: Enterprise_Architecture, Engineering_View, Telco.

1 Introduction

1.1 Two Challenges for Telcos

In this paper, we will present two examples of how Enterprise Architecture could help make better investment decisions within a multi-country Telecommunications

---

Hervé Pacault
Orange Labs, Paris, France
e-mail: herve.pacault@orange.com

P.-J. Benghozi, D. Krob, & F. Rowe (Eds.): *Digital Enterp. Des. & Manage. 2013*, AISC 205, pp. 23–32.
DOI: 10.1007/978-3-642-37317-6_3

Company. These two examples relate to two of the significant challenges that a Telco is currently facing :

- The invasion of Telco's traditional playground by new actors, the web players and the consumer electronics manufacturers
- The sharing of IT components between the local companies of the same Telco Group, in order to cut cost through economies of scale

For each of the examples, we will present a strategic view and an engineering view, that could be seen as

- a simplified version of the application view in the first example,
- and a simplified infrastructure view in the second example.

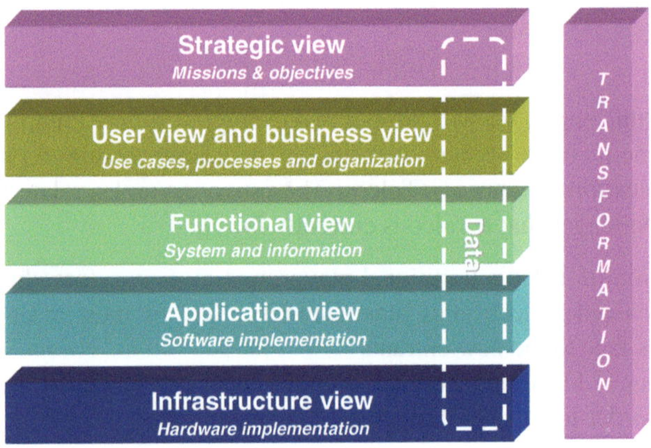

Fig. 1 Enterprise Architecture framework (source Club Urba-EA)

In both cases, the objectives of the Enterprise Architecture modeling is to facilitate the dialog between various stakeholders within the company : Strategy, Marketing, and IT, in order to get a consensus on how fast, how far should transformation take place within the company when

- Shifting from a closed model to an open model, in the first example
- Setting data centers at the Region level, and populating them with IT components shared by the Telco operating companies in the different countries, in the second example

1.2 Modeling Conventions

When modeling, we will try, as much as possible, to refer to the TM Forum (TeleManagement Forum).

TM Forum is a global, non-profit industry association focused on enabling service provider agility and innovation. As an established thought-leader in service creation, management and delivery, the Forum serves as a unifying force across industries, enabling more than 900 member companies to solve critical business issues through access to a wealth of knowledge, intellectual capital and standards.

Fig. 2 TM Forum board members

TM Forum Provides Several Frameworks Such as

- Business Process Framework (eTOM)
- Information Framework (SID)
- Application Framework (TAM)
- Integration Framework

What is specific about a Telecommunication Company, or Telco, is that *the business activity of the company consists almost only in transporting, transforming and storing data.* With the digitalization of voice, and the preeminence of IP, what is now being delivered to the customer consists mostly in data. The share of other products and services sold to the customer (handset and equipment, training, financial or insurance services etc) in the sales of a Telco is marginal.

Therefore, *the starting point for modeling will be the processes between the customer and the Telco*, because these processes embody almost the entirety of the billable activity of a Telco. In this document, we will split these processes in two sets, that we will refer to under simplified wordings :

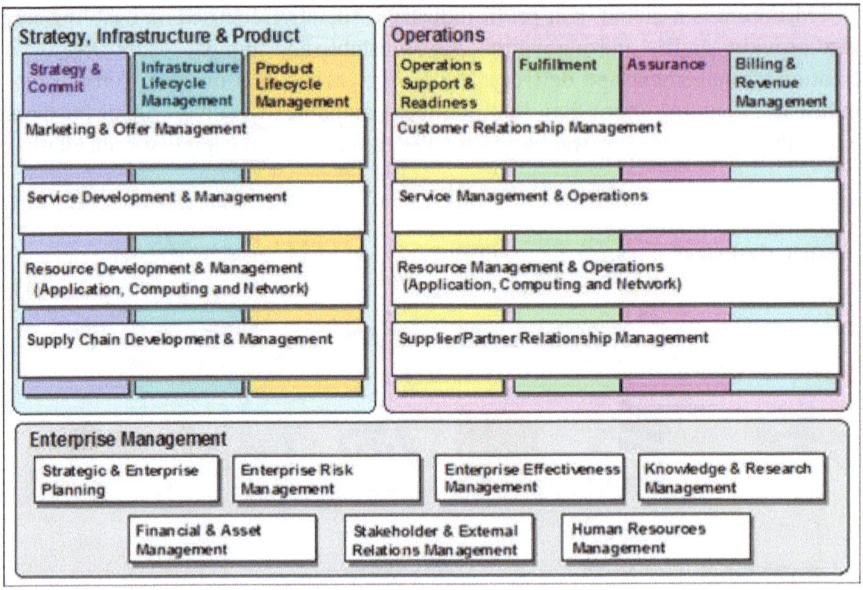

Fig. 3 eTOM

- *"Subscription, billing and ecare"* workflows between the Telco and the "Customer", as shown on the eTOM map below[1].
- *"Service Usage"* workflows between the Telco and the "User" : The corresponding processes are not yet explicitly described in eTOM

For the *application view*, we will use the acronyms *"OSS & BSS"*, because these two acronyms are well understood in the IT world[2], in lieu of the domain detailed classification from TM Forum.

Applications that directly deliver services consumed by the user will be represented in this paper under the name "*User Service*". They are not yet explicitly classified in TMF Application Framework classification, that currently focuses on the applications that are the day-to-day tools of the employees of the company.

---

[1] The Business Process Framework (eTOM) is a widely deployed and accepted model and framework for business processes in the Information, Communications, and Entertainment industries. The Business Process Framework represents the whole of a Service Provider's enterprise environment.

[2] Operations support systems (OSS) are computer systems used by telecommunications service providers. The term OSS most frequently describes "network systems" dealing with the telecom network itself, supporting processes such as maintaining network inventory, provisioning services, configuring network components, and managing faults. The complementary term Business Support Systems (BSS) refers to "business systems" dealing with customers, supporting processes such as taking orders, processing bills, and collecting payments. The two systems together are often abbreviated OSS/BSS, BSS/OSS or simply B/OSS. (quoted from Wikipedia).

2 First Example : The Invasion of Telco's Traditional Playground by New Actors, the Web Players and the Consumer Electronics Manufacturers

2.1 Telco versus Web Player : Strategic View

Many of the services that are consumed by Telco customers on their smartphone, TV set, tablet or laptop are already produced by third parties.

Thanks to technology progress, such as ip and http, more and more of the services that a Telco is delivering, a web player also could offer, or would be able to offer, even if it is with a somehow different Quality of Service.

A Telco must then decide to which extend it should compete against these web players, and to which extend it could benefit from partnering with them. For Telcos, the objective would be to develop the business with these web players under partnership mode, by exposing assets such as network integration, Identity or Billing, and not to limit the Telco's offering to pure IP.

Such competition / partnership decision belongs to Strategy, but has also deep implications in IT architecture.

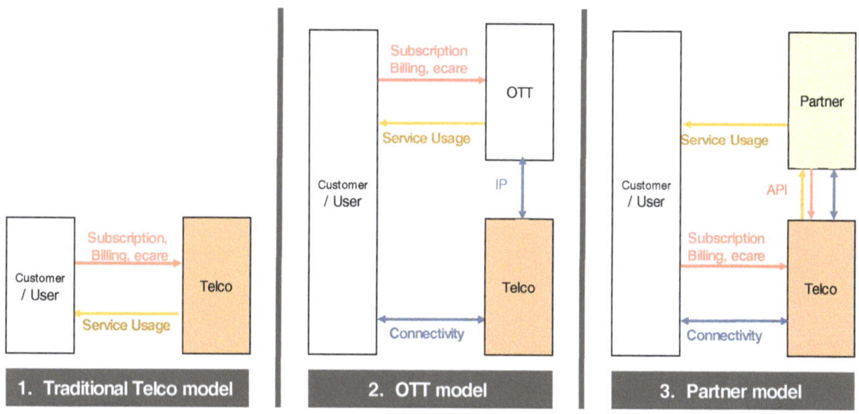

Fig. 4 Web player Strategic view

Over-The-Top players (OTTs) only consume pure IP from Telcos. They build their own Subscription, Billing and Identity links with the customer. In some cases, they could directly trigger Telco's services on behalf of the customer.

Telco's objective is to provide high value added services to the Service Providers, exposing their assets through APIs, along Service Level Agreements and, as much as possible, keep control of the Subscription, Billing and Identity of the customer.

2.2 *Telco versus Web Player : Engineering View*

In the *OTT model*, "usage" can be viewed as the combination of two flows :

- IP connectivity
- Service usage (voice call, messaging, TV and VOD, payment etc), that relies upon IP connectivity

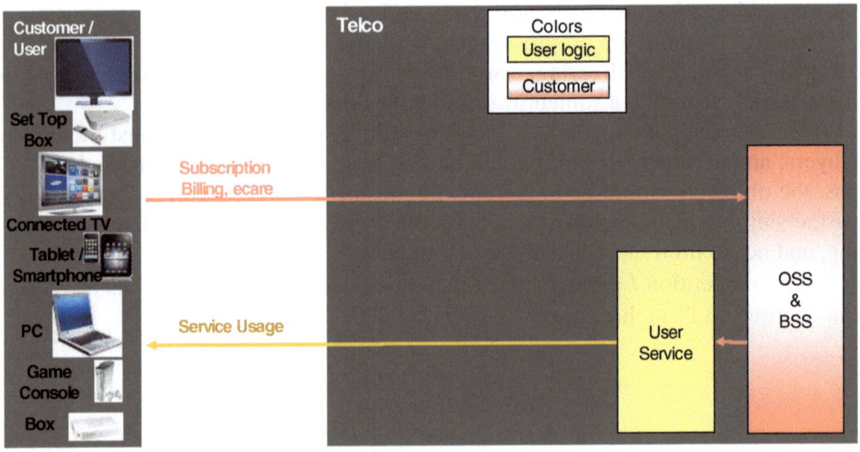

Fig. 5 Telco Traditional model

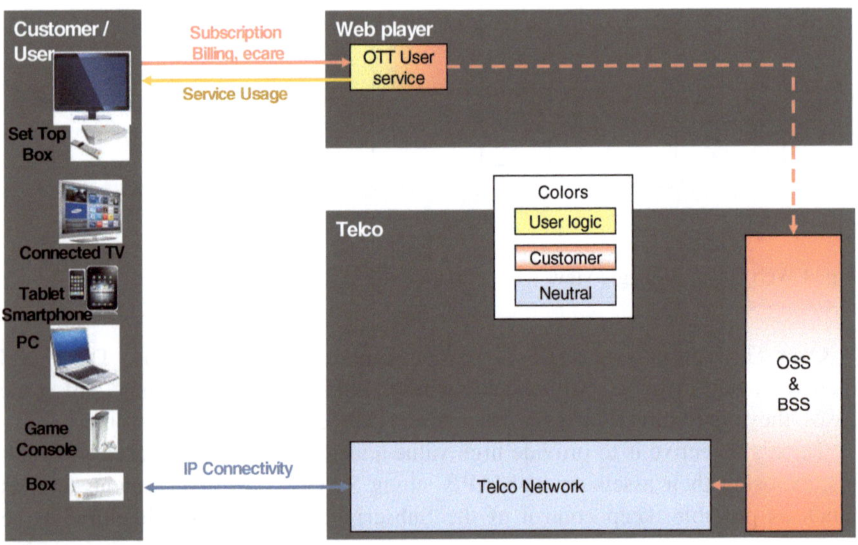

Fig. 6 Over-The-Top model

In the *Partner model,* in order to ensure seamless navigation between the Telco services and the services offered by the different Web players, the Telco will try to keep the management of the authentication / identification of the customer. Customer will then benefit from Single-Sign-On. This Control function (user authentication, session control, access rights management, sometimes summarized under the name of Identity) could also, in our representation, be set apart from "Service Usage".

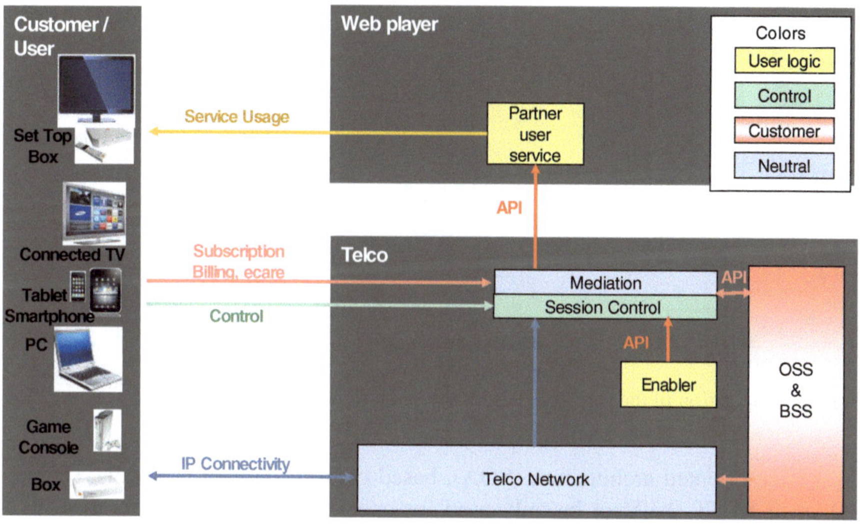

Fig. 7 Partner model

The exposition of the Telco's assets (network, billing, identity, enablers) to the Web player is being made possible by technologies such as Service-oriented architecture (SOA), based on well defined protocols, such as SOAP or REST, that

- transform capabilities into service interfaces and publishes them
- define, expose and control high level interfaces (the APIs), making abstraction of underlying physical resources

All components (user logic components, enablers, OSS/BSS) are connected to a mediation layer through APIs[3]

We can now represent the three models on the same figure :

[3] **Mediation** : An API mediation system is an essential infrastructure for the deployment of Service Orient Architecture. It may embed functions such as : Exposition of basic services, Orchestration of basic services in order to deliver enhanced services, Brokering between providers and consumers of APIs, Management of requests made by consumers of APIs, both for internal and external APIs consumers.

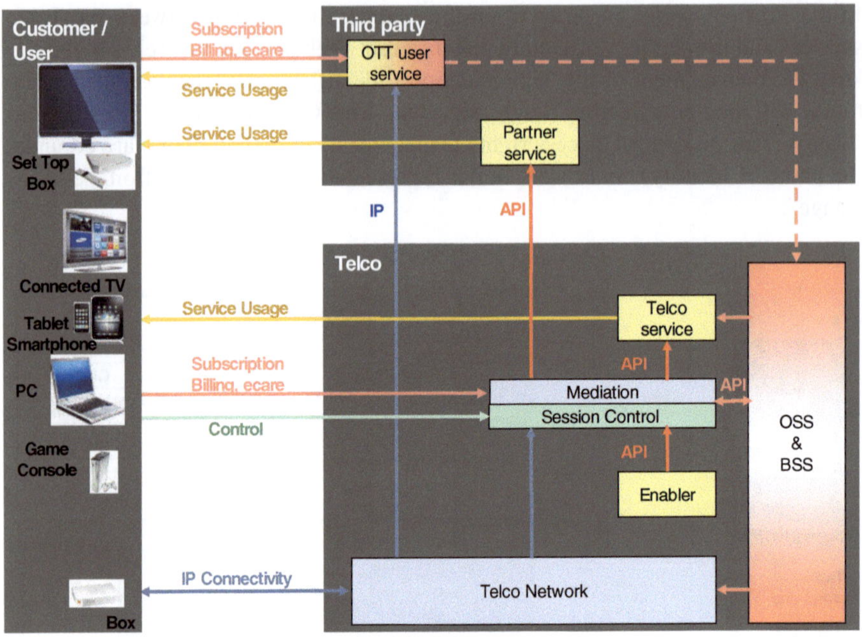

Fig. 8 Juxtaposition of the three models on the same figure

Service-oriented architecture (SOA), based on well defined protocols, such as SOAP or REST shall not be only used for opening the Telco's assets to a third parties, but will also be used by the Telco for the integration of the services that it provides to its customers since it

- Improves flexibility, the re-use of components and Quality-of-Service
- And therefore reduces overall build and run cost.

3 Second Example : Sharing IT Components between Different Countries

3.1 Multi-country Telco : Strategic View

An international Telco usually has operations and customers in a large number of countries, often over 20. Each of these operating companies is facing different local market conditions, and is constrained by a different IT legacy. Therefore, a large autonomy in IT investment decisions is usually given to the local operating company.

However a large share of the IT processing could be shared between several countries. This would bring significant benefits : cut down cost, facilitate the deployment of new services, facilitate consistent customer experience between countries etc. We all know that Web players architecture is totally centralized, with very limited local implementation. So, why not do the same for Telco's architecture?

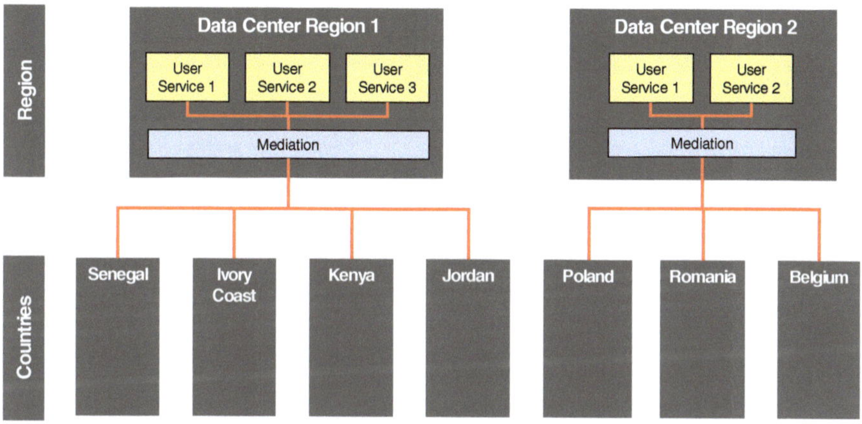

Fig. 9 Multi-country strategic view

Technical limits to this IT sharing could stem from

- Bandwidth constraints (lack of international IP backbone for instance in some African countries, or large data volumes for instance for TV services)
- Response time constraints (such as with voice telephony services or voice and data charging)

3.2 *Multi-country Telco : Engineering View*

On the below figure, we can see that the applications delivering the User Services could be instantiated in four different locations :

- *Spread in the country* : This was the case of Central Switches delivering traditional voice telephony services, before the all IP era.
- *Telco centralized in the country* : This is the most frequent. In this category one could find Voice over IP services, SMS, TV and services that have high bandwidth or specific response time requirements
- *Telco shared in Group or Region* : This is the trend that Telcos have been following with http services such as Messaging, Payment and Business services, and other value-added-services
- *Web player* : As mentioned in the first example, customers consume more and more value-added-services provided by web players

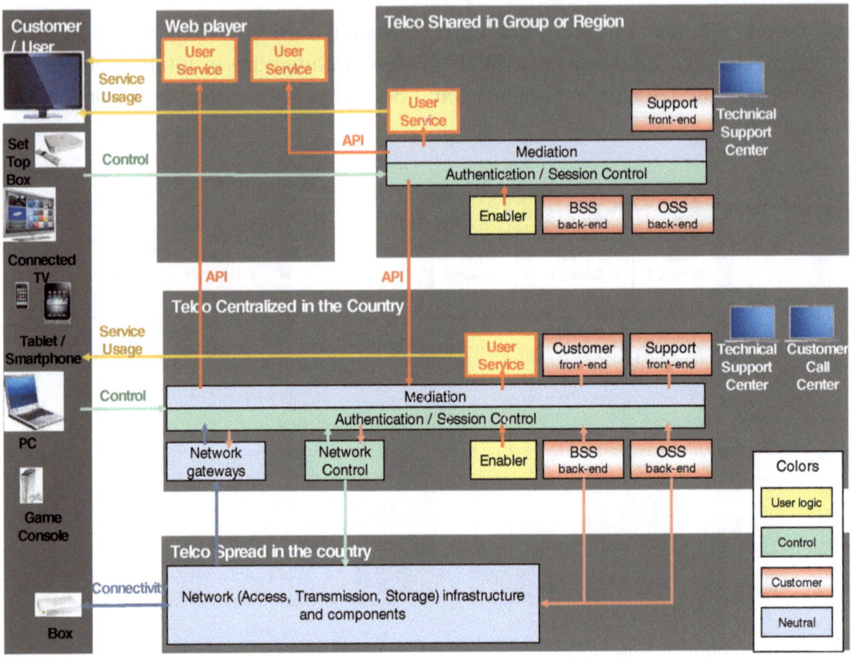

Fig. 10 Multi-country Telco : Engineering view

With these graphic conventions, *it is possible to show, on a single A4 sheet, a summary of the technical policy* of the Telco regarding partnership and instantiation choices for the delivery of end-user services.

4 Conclusion

These graphic representations are none of the traditional views used by architects in a project (which are : use view, functional view, application view, infrastructure view), but rather a combination of application and infrastructure views. We believe that the above-demonstrated views will be easier to grasp by non-architects, and will therefore could enrich the dialog between all stakeholders within the company, making it easier to align IT on business.

References

`http://www.tmforum.org/`
`http://www.urba-ea.org/`
`http://www.opengroup.org/togaf/`

GrammAds: Keyword and Ad Creative Generator for Online Advertising Campaigns

Stamatina Thomaidou, Konstantinos Leymonis, and Michalis Vazirgiannis

Abstract. Online advertising is a fast developing industry - in 2011 its revenues reached $31 billion. Paid search marketing is extremely competitive while online advertising campaign creation and development are very demanding in terms of time and expert human resources. Assisting or even automating the work of an advertising specialist has emerged as a requirement for companies and research institutes over the last few years, mainly because of the commercial value of this endeavour. In this context, we developed *GrammAds*, an automated keyword and ad creative generator. This system generates multiword keywords (n-grams) and automated ad creative recommendations, while it organizes properly the campaigns which are finally uploaded to the auctioneer platform and start running. In this paper, we analyze the proposed methodology and we also present the main functionality of the *GrammAds application* along with an experimental evaluation.

1 Introduction

Online advertising is gaining acceptance and market share while it has evolved into a $31 billion industry for advertisers[1]. One form of online advertising is the promotion of products and services through search-based advertising. The three most prevalent options in the search-based advertising market are Google AdWords,

Stamatina Thomaidou · Konstantinos Leymonis
Athens University of Economics and Business
e-mail: thomaidous@aueb.gr, k.leymonis@dias.aueb.gr

Michalis Vazirgiannis
LIX, Ecole Polytechnique & Athens University of Economics and Business
e-mail: mvazirg@lix.polytechnique.fr

[1] http://www.iab.net/AdRevenueReport

P.-J. Benghozi, D. Krob, & F. Rowe (Eds.): *Digital Enterp. Des. & Manage. 2013*, AISC 205, pp. 33–44.
DOI: 10.1007/978-3-642-37317-6_4

Yahoo Search Marketing, and Microsoft AdCenter (the two latter have merged)[2]. Today's most popular search-based advertising platform is Google AdWords having the largest share of revenues amongst its competitors. Search remains the largest online advertising revenue format, accounting for 46.5% of 2011 revenues, up from 44.8% in 2010. In 2011, Search revenues totaled $14.8 billion, up almost 27% from $11.7 billion in 2010. Search has remained the leading format since 2006, having strong sequential growth.[3]

Baseline Campaign Creation Process: To create an advertising campaign usually the advertiser must have one or more products on a website that wants to be exposed to the public. For each product there must be a landing page, which is the webpage a user will be redirected to, after clicking the advertisement of the product. The landing page is usually the place where the user can see information about the product, its technical characteristics, its price, and has the option to buy it. After finding what the advertising wants to sell and preparing the landing pages, it is crucial to select on which keywords (words or phrases) each product will be advertised. The keywords used for each product must be relevant to it, otherwise the campaign will not be profitable. A good practice is to choose the most specific key-phrases possible, which usually consist of 1-3 words.

After finding the keywords, the advertisement texts (or ad creatives) must be correctly written. They must be short and precise, understandable with convincing calls to the user to take action. Ad creatives consist of a short headline, two limited lines of description and a display url, which does not have to be the same as the real landing page url. Keywords and ads belong to AdGroups. The ads of an AdGroup are shown for keywords belonging to the same AdGroup. Thus, it is important not to mix ad-texts and keywords of irrelevant products in the same AdGroup. A budget must also be set on every campaign which will be consumed. An advertiser must also decide how much the maximum cost-per-click of each keyword will be. This is the bid that the advertiser is putting for a keyword and approximately an upper limit of how much each click for this keyword may cost.

Motivation for an Automated Application: As it is implied, effective *keyword selection* is one of the most important success factors for online advertising. Companies would like to advertise on the most effective keywords to attract only prospective customers and not uninterested browsing users, while they are also in need of well-written ad creatives to attract more visitors and generate thus higher revenues. In addition, the preparation of large scale online advertising campaigns for products, services, brands, or web pages can be a very complex task if it is designed for websites with *online catalogs* or catalog aggregators (e.g. `http://www.fnac.fr/`). The shops or listings are classified according to the products that they are selling, so each landing page contains important information and relevant description for each category or product that needs to be considered. The number of the various urls inside these domains makes the effort even more complicated regarding the manual

[2] `http://www.searchalliance.com/publishers`

[3] `http://www.iab.net/AdRevenueReport`

insertion of keywords and ad-texts per landing page. Our proposed system aims at the automation of the total procedure in order to aid the advertisers.

2 Related Work

The manual selection of even a small set of keywords for advertising purposes is quite laborious, a fact which leads to the recent appearance of commercial tools that create keyword sets directly from a landing page. There exist different techniques for keyword generation. Search engines use query log based mining tools to generate keyword suggestions. In this way, they focus on discovering co-occurrence relationships between terms and suggest similar keywords. They start from an initial key phrase and they are based on past queries that contain these search terms. Google AdWords Keyword Tool[4] exploits this ability and presents frequent queries for the seed set of terms

Other commercial tools[5] determine an advertiser's top competitors and then actively search for the keywords they are targeting. After a period of time, lists of targeted keywords that are competitive for pay per click advertising are automatically generated. These two approaches may result to a recommendation set of keywords which are likely to be general and thus more expensive. Considering this, the challenge of generating keywords is to select both semantically similar and well-focused keywords.

TermsNet and Wordy [8, 1] exploit the power of search engines to generate a huge portfolio of terms and to establish the relevance between them. After selecting the most salient terms of the advertiser's web page they query search engines with each initial seed term. With their methods they find other semantically similar terms. Wordy system proposed single word terms (unigrams) for each seed keyword. S. Ravi et al. [12] propose a generative model within a machine translation framework so the system translates any given landing page into relevant bid phrases. They first construct a parallel corpus from a given set of bid phrases b, aligned to landing page keywords l, and then learn the translation model to estimate $Pr(l|b)$ for unseen (b,l) pairs. This approach performs very efficiently but depends on the chosen domain and data that the human decision factor may affect.

In general, *corpus or domain dependent* systems require a large stack of documents and predetermined keywords to build a prediction model [11], while on the other hand our developed system works with a *corpus independent* approach that directly sifts keywords from a single document without any previous or background information.

Regarding the automated ad creative generation process, to the best of our knowledge, this issue remains still an open problem in Natural Language Processing and Information Retrieval areas as mentioned in [6]. Thus, the corresponding module of our system is an innovative contribution in this regard.

---

4 `http://www.adwords.google.com/keywordtool`

5 `http://www.adgooroo.com/`, `http://www.wordstream.com/`

3 Keyword Generation

This component aims at proposing valid and representative keywords for a landing page capitalizing on keyword extraction methods, on the co-occurrence of terms, and on keyword suggestions extracted from relevant search result snippets.[6] A more detailed presentation of this component is described in [14]. In the following two paragraphs we will describe briefly their main functionality.

Keyword Extraction: In this process, we follow the corpus independent approach to rely solely on the given landing page document. As a preprocessing step, the HTML content of each landing page is parsed, stopwords are removed and the text content is tokenized. Next, for each word (unigram) in the tokenized output, we compute a special tag weight. From the most retrieved unigrams we pull together possible combinations of two-word phrases (bigrams) inside the given landing page. Following the same co-occurence discovery process we extract three-word terms (trigrams) as well. By gathering all terms, we construct the final extracted keywords vector. We boost trigrams first, bigrams second and unigrams third, modifying their relevance score proportional to the number of grams.

Keyword Suggestion: From the previous step of keyword extraction we have already extracted the initial keywords. These will be the seed keywords for the additional suggestions For each given seed keyword, the keyword is submitted as a query into a search engine API. The top 30 snippet results are downloaded and loaded in Apache Lucene Library[7] as small documents. We parse them and we construct a new vector of grams. Based on the Lucene scoring method we find the unigrams and bigrams that have the largest number of occurrences inside the document and thus are kept as the most relevant for the specific seed query. Each of these terms is representing a new query. Again, we sort in descent order the new queries based on this score and we create a vector of suggested keywords and their scores for each of the seed terms.

4 Ad Creative Generation

Here, we propose an automated process for the *ad creative generation*. Summaries of Web sites help Web users get an idea of the site contents without having to spend time browsing the sites. The technology of automatic text summarization is maturing and may provide a solution to the information overload problem. Automatic text summarization produces a concise summary of source documents. The summary can either be a *generic summary* (this type of summary we using in the next process), which shows the main topics and key contents covered in the source text, or a *query-relevant* summary (which is a further challenge of our system as we could use specific keywords for filtering the resulted summaries), which locates the contents

[6] A demonstration of the process can be found in the developed web application at `www.grammads.com`

[7] `http://lucene.apache.org/java/docs/index.html`

pertinent to user's seeking goals [15]. In this subprocess the first step was to extract all the text from the HTML document of the given landing page. Then, we used summarization to keep the most important meaning for the description of our advertising page. For this purpose the input was the text from the page to the Classifier4J[8]. Using a Bayesian classifier implementation, we can get a text summarization of one sentence -which is considered as the most important one retrieved from the landing page- in order to insert it into the description lines of the ad creative. In the end of the second description line we add a call-to-action phrase such as: "Buy now!", "Purchase now!", "Order now!", "Browse now!", "Be informed" according to each advertising goal. For example we use "Be informed" for advertising a web page or a brand name, "Buy now!" for advertising a product, "Purchase now!" for advertising a service. The purpose of adding these phrases into the text ad is to optimize the ad descriptions because a call-to-action encourages users to click on the ad and ensures they understand exactly what the advertiser expects them to do when they reach the landing page.

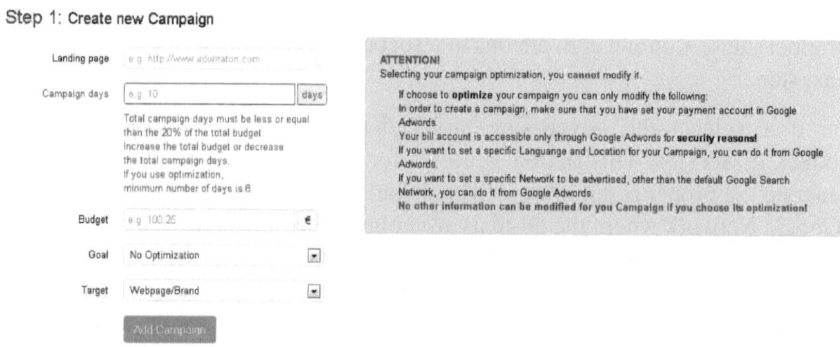

Fig. 1 Initialization Settings

5 Campaign Organization and Use Cases

The main objective of the system developed is to improve the user experience of setting up and optimizing small or large scale Google AdWords Campaigns, no matter how experienced the user is, in a very simple, straight-forward way, minimizing the needed input information and the total steps of the process. In that way, GrammAds application significantly reduces time, resources, and final effort of the advertisers. For a mid-size one product campaign (i.e. 3 Adgroups - 30 keywords) the labour cost for a manual campaign is $\sim$ 30 person hours per month, whereas using the proposed system the whole process drops to a few minutes for inserting the desired landing pages and the basic goals of the campaign.

[8] `http://classifier4j.sourceforge.net/`

In the following paragraphs we will present a basic example of the application use. In each case, user has to have been logged in the system, with the AdWords credentials. Adomaton Servlet starts a new session for this user, assigning all the information needed between requests and responses, to an object, which then is being saved in this specific session as attribute. User selects creating a new AdWords campaign and he is directed to the page that is presented in Figure 1. User fills the information needed for the campaign that the application is going to create, inserts the main URL of the product, service, or brand-name that he wants to promote, the period days of the whole campaign, and the budget for the total campaign days. In addition, he selects one of the three system runnable options for the campaign that is going to be activated:

1. *No Optimization*, where the system just uploads automatically the generated keywords, ad-texts, and bids along with their organized structure without continuing to be responsible for an automated optimization strategy
2. *Traffic Optimization*, where the advertiser considers the profit to be the amount of clicks at the ad-texts
3. *Profit Optimization*, where the profit is the actual monetary profit from offline product sales or online conversions to a specific landing page that is defined in a next step

Then, user selects the advertising target that can be a.Website/Brand-name, b.Product, c. Service. This option is useful for the Ad Creative Module in order to generate the proper action phrase. The default option from our system regarding the Google Network where the ad-texts are going to be impressed is only the group of *Search Network*, opting-out the same time from the Display Network group. We took this decision in our experiments and strategies because choosing to appear also in the Display Network was leading to a large amount of impressions and very few clicks. As a result the values of CTR (Clickthrough rate) were very low ($< 0.5\%$) causing in this way low Quality Scores and increased recommended bids for good ad slots.

After the proper settings are inserted, Adomaton Servlet as a second step reads each input information and assigns them to the session object. Depending on the main landing page that has been set by the user, the *Crawler Module* visits the specified webpage and using a web-scraping technique, obtains webpage source information. In this process, the Crawler extracts possible existing sub-landing pages, after validating their availability. Each retrieved sub-landing page is corresponding semantically to an AdGroup in our proposed Campaign Organization. For the case of Profit Optimization, next to each sub-landing page the user can insert a specific monetary profit that he is going to gain from a conversion in this page. This is useful for the Profit Maximization Strategy (a detailed algorithm is presented in [10]). The user can select all or a portion of these retrieved landing pages.

As a third step, the user is directed in a page where he must select for each of the previously selected pages the automatically generated keywords. Next to each keyword it is presented to the user a normalized score of its relevance to the AdGroup, as well as an initial bid value. This value is derived from

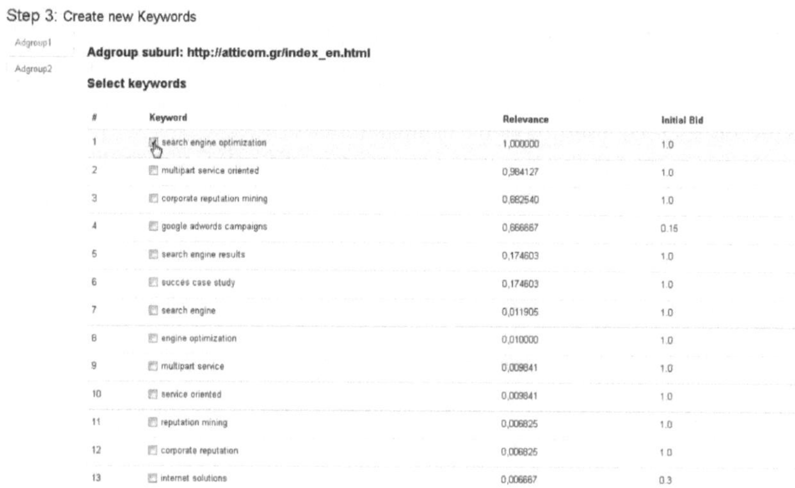

#	Keyword	Relevance	Initial Bid
1	search engine optimization	1,000000	1.0
2	multipart service oriented	0,984127	1.0
3	corporate reputation mining	0,682540	1.0
4	google adwords campaigns	0,666667	0.15
5	search engine results	0,174603	1.0
6	succes case study	0,174603	1.0
7	search engine	0,011905	1.0
8	engine optimization	0,010000	1.0
9	multipart service	0,009841	1.0
10	service oriented	0,009841	1.0
11	reputation mining	0,006825	1.0
12	corporate reputation	0,006825	1.0
13	internet solutions	0,006667	0.3

Fig. 2 Automatically generated keywords and recommended bids

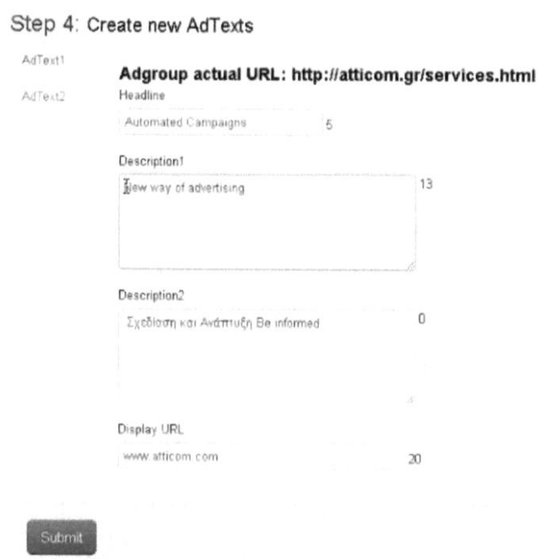

Fig. 3 Automated Ad Creatives

$min(1, estimatedFirstPageCPC)$. This step is presented in Figure 2. The estimated first page bid amount approximates the cost-per-click (CPC) bid needed for the ad to reach the first page of Google search results when a search query exactly matches the keyword. This estimate is based on the Quality Score and current advertiser competition for that keyword. We retrieve this value using the AdWords API[9]. We decided to place an upper bound of 1 due to the fact that a commonly used strategy

[9] `https://developers.google.com/adwords/api/`

by the advertisers is to evaluate as a default case the utility of the click at a url to be equal to one monetary unit (e.g. 1 euro or dollar) [4, 2].[10]

Furthermore, the user at the fourth step in Figure 3 can select for each sub-landing page the automatically generated advertising text as well as edit each part of it to his needs, before its final insertion in the campaign.

Finally, in Figure 4 the user can see through the AdWords interface the uploaded settings of his constructed campaign.

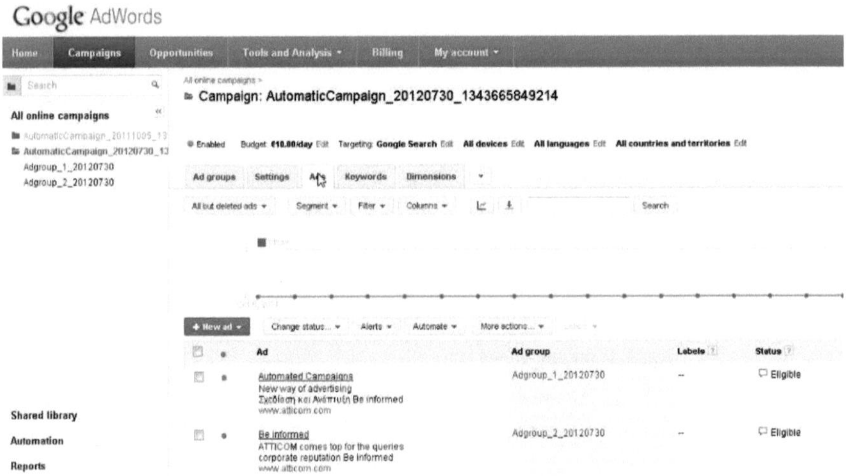

Fig. 4 AdWords Uploaded Settings

6 System Architecture and Communication

The proposed system structure resulted as a need for an appropriate user interface interaction of the *Adomaton* subsystem, which optimizes Google AdWords Campaigns through a novel bidding strategy. In this logic, it is intended to allow users access the system through a web interface, available from their browsers, just by logging in a webpage, using their registered account's AdWords credentials. Each registered Google AdWords user, can use the system as a client. Our web application is equipped with its own server and database, in order to keep the needed information such as statistics and perform the initial upload of the campaign in the corresponding user account on AdWords API, optimize and finally monitor the campaign's performance.

For the development of the interaction system, Servlet is used, a Java-based server-side web technology, which extends the capabilities of a server that host applications access via a request-response programming model. This technology is

[10] `http://support.google.com/adwords/bin/answer.py?hl=en&answer=2471184&from=50081&rd=1`

tied with the HTTP protocol in order to comply with the web browser access to it. The main reason that Servlet has been used is because of the optimization algorithms that have been previously developed and tested with the Java programming language. In order to display the content of the information Model-View-Controller (MVC) is used as a pattern for the information management. More specifically the Java Server Pages (JSP) technology were included in order to dynamically generate web pages based on HTML markup notation language. To deploy and run Java Server Pages , it was required Apache Tomcat, a compatible web server with a Servlet container. AdWords API has been used in order to ensure the connection of the system with the AdWords system, and thus use the AdWords modules to synchronize each campaign's options and settings with our system. The development methodology that has been used is the incremental development [9], a combined linear-iterative framework. In that way, it was performed a series of waterfalls consisting of requirements extraction, design, implementation, testing and maintaining of the system. An important aspect that we took under consideration are the constant AdWords API changes and migrations. Thus, in need of a more proper code

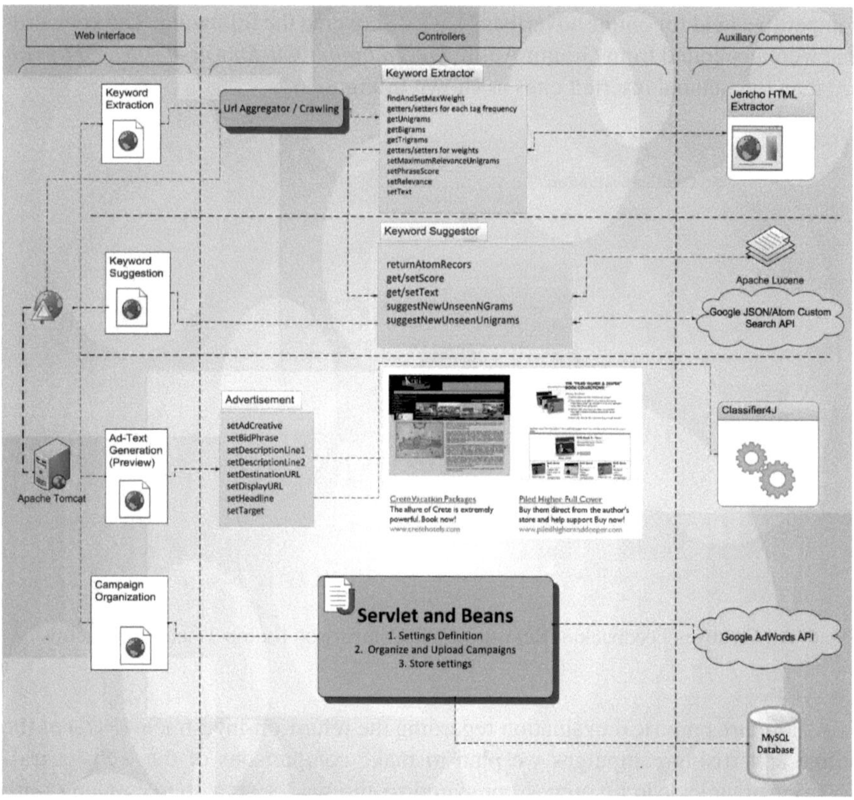

Fig. 5 System Architecture and Modules Communication

organization and maintenance, we constructed a package of *AdWords API Wrapper*, to keep up with all the corresponding library major edits.

In Figure 5, we present the *GrammAds* architecture with the associations of the basic modules. This system was developed in the context of an *overall automated solution* for creating, monitoring, and optimizing a Google AdWords campaign. A detailed description of the *Budget Optimization* process and main strategy can be found in [10], as well as a beta demo version in [13].

7 Experimental Evaluation

In our initial experiments described in [14], we evaluated at a first glance the recommended keywords using human ranking following a blind testing protocol. In the next steps we evaluated also the automatically generated ad-texts. Eleven researchers and postgraduate students provided feedback in the scale of Grade.1:Bad up to Grade.5:Very Good. In Figure 6 we present the evaluation results. In the context of various experiments where we used the exact same bidding strategy for two identical campaigns of a company that offers web developing solutions (a highly competitive field for online advertising), we discovered the following: The keywords that were generated from GrammAds *achieved higher Clickthrough rate (CTR) values* than the manual inserted ones as shown in Figure 6.

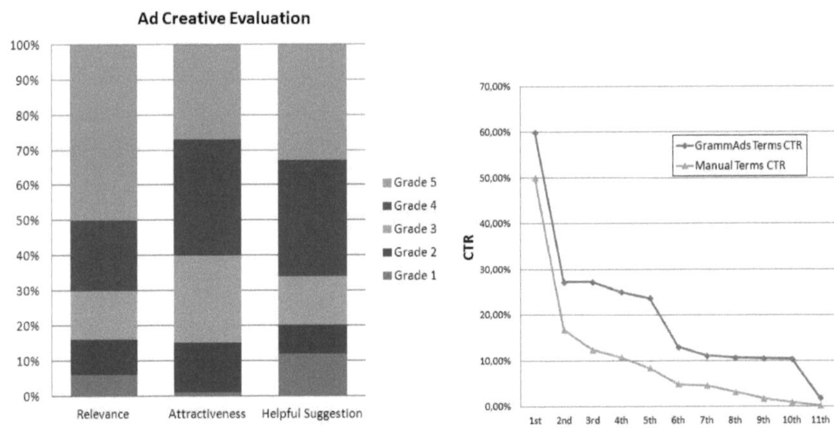

Fig. 6 Ad Creatives Feedback & Keyword CTR Comparison for top-11 generated terms

As a future empirical evaluation regarding the return-on-investment (ROI) of the online advertising campaigns we plan to make comparisons of the website traffic from organic-search sources compared to the paid-search sources along with their matched search query[11]. In order to make an evaluation more focused on the

[11] `http://www.google.com/analytics/`

quality of the recommended keywords and ad creatives by the GrammAds application rather than focusing on more complex and sophisticated bidding strategies, we will set each day as a bid value the recommended initial bid value of our system ($min(1, estimatedFirstPageCPC)$).

8 Conclusions and Future Work

In this paper we proposed a system that, given a landing page in the context of online advertising for products and services promotion, automatically extracts and suggests keywords for web advertising campaigns as well as automatically generates advertisement texts.

In this way, our contributions regarding the improvement of the advertising campaign creation process consist in:

- Automating the task of finding the appropriate keywords
- Recommending multiword terms (n-grams) with high specificity without the need to capitalize on usage data such as query and web traffic logs
- Generating fast snippets of ad texts
- An automated and fast way of uploading campaign and AdGroup settings (keywords and ad-texts per AdGroup) into the AdWords service
- A developed web application with an initial experimentation on real world campaigns

Using the search result snippets for the process of keyword suggestion has helped a lot to retrieve faster the proper information rather than crawling actual documents. It was also a helpful mean to keep the trends and thus retrieving trending topics at a specific time. Also, searching result snippets from queries on twitter search and tags can be helpful due to the compact nature of twitter messages. They can help in filtering out irrelevant or general information, while mining market trends. A further extension on our system can be the expansion of the ad creative generation component. The creation of specialized ad text will be based on previous work and research studies on paraphrasing methods, sentence extraction and compression, sentence and surface realizers, and text summarization [3, 5, 15, 7]. As a future challenge for more attractive advertisements, the system could take into account sentiment analysis as well. Further steps of our future work regarding expansion of user requirements from our application are: a. Improving task scheduler and take into consideration the possible concurrent use of the system from a large volume of new users. b. Allowing users to select if they want to construct a completely new campaign or to optimize an already running one, regarding bidding adjustments of existing keywords. For the latter, our system will migrate the information that is being kept in our database tables to the existing settings of the AdWords Campaign.

Acknowledgements. The research of S. Thomaidou is co-financed by the European Union (ESF) and Greek national funds via the NSRF Program: Heracleitus II. Prof. M. Vazirgiannis is partially supported by the DIGITEO Chair grant LEVETONE in France.

References

1. Abhishek, V., Hosanagar, K.: Keyword generation for search engine advertising using semantic similarity between terms. In: ICEC, pp. 89–94 (2007)
2. Abhishek, V., Hosanagar, K.: Optimal bidding in multi-item multi-slot sponsored search auctions. In: EC 2012 (2012)
3. Androutsopoulos, I., Malakasiotis, P.: A survey of paraphrasing and textual entailment methods. J. Artif. Intell. Res. (JAIR) 38, 135–187 (2010)
4. Borgs, C., Chayes, J., Immorlica, N., Jain, K., Etesami, O., Mahdian, M.: Dynamics of bid optimization in online advertisement auctions. In: WWW 2007, pp. 531–540 (2007)
5. Choi, Y., Fontoura, M., Gabrilovich, E., Josifovski, V., Mediano, M., Pang, B.: Using landing pages for sponsored search ad selection. In: Proceedings of the 19th International Conference on World Wide Web, WWW 2010, pp. 251–260. ACM, New York (2010)
6. Gabrilovich, E.: Ad retrieval systems *in vitro* and *in vivo*: Knowledge-based approaches to computational advertising. In: Clough, P., Foley, C., Gurrin, C., Jones, G.J.F., Kraaij, W., Lee, H., Mudoch, V. (eds.) ECIR 2011. LNCS, vol. 6611, pp. 4–5. Springer, Heidelberg (2011)
7. Galanis, D., Androutsopoulos, I.: An extractive supervised two-stage method for sentence compression. In: HLT-NAACL, pp. 885–893 (2010)
8. Joshi, A., Motwani, R.: Keyword generation for search engine advertising. In: ICDM Workshops, pp. 490–496 (2006)
9. Larman, C., Basili, V.: Iterative and incremental developments. a brief history. Computer 36(6), 47–56 (2003)
10. Liakopoulos, K., Thomaidou, S., Vazirgiannis, M.: The adomaton prototype: Automated online advertising campaign monitoring and optimization. In: 8th Ad Auctions Workshop, EC 2012 (2012)
11. Liu, J., Wang, C., Liu, Z., Yao, W.: Advertising keywords extraction from web pages. In: Wang, F.L., Gong, Z., Luo, X., Lei, J. (eds.) WISM 2010. LNCS, vol. 6318, pp. 336–343. Springer, Heidelberg (2010)
12. Ravi, S., Broder, A.Z., Gabrilovich, E., Josifovski, V., Pandey, S., Pang, B.: Automatic generation of bid phrases for online advertising. In: WSDM, pp. 341–350 (2010)
13. Thomaidou, S., Leymonis, K., Liakopoulos, K., Vazirgiannis, M.: Ad-mad: Integrated system for automated development and optimization of online advertising campaigns. Demo to appear in ICDM 2012 Workshop Proceedings (2012)
14. Thomaidou, S., Vazirgiannis, M.: Multiword keyword recommendation system for online advertising. In: ASONAM 2011, pp. 423–427 (2011)
15. Zhang, Y., Zincir-Heywood, N., Milios, E.: World wide web site summarization. Web Intelli. and Agent Sys. 2, 39–53 (2004)

Interoperable Systems and Software Evolution: Issues and Approaches

Norman Wilde, Sikha Bagui, John Coffey, Eman El-Sheikh, Thomas Reichherzer, Laura White, George Goehring, Chris Terry, and Arthur Baskin

Abstract. Interoperability is essential for modern enterprise software; one of the most promising ways of providing interoperability is though Services Oriented Architectures (SOA) usually implemented using the Web Services (WS) standards. SOA/WS has the potential to be a transformational technology but there are a number of problems that may hinder its application. One of these is the classic slowness of software evolution. This paper discusses the issues of SOA evolution and describes ongoing research experimenting with the use of search technology to speed comprehension of SOA applications. Flexible but specialized search tools may be a good match for the "open world" of a SOA system which may encounter frequent novelties in programming languages and technology during its lifetime. We describe a basic search tool adapted to SOA/WS artifacts, a knowledge-based extension to it to improve software comprehension, and ongoing work to handle additional document types and to provide ontology-based support. Development of support tools for SOA evolution could be a fruitful topic for industry-university collaboration. Such tools would be an enabler for the interoperable information systems needed to do business in the modern world.

1 Introduction

Two trends in business and government drive the growing need for interoperable information systems:

Norman Wilde · Sikha Bagui · John Coffey · Eman El-Sheikh · Thomas Reichherzer ·
Laura White · George Goehring · Chris Terry
University of West Florida, Pensacola, Florida, U.S.A.
e-mail: {nwilde,bagui,jcoffey,eelsheikh,treichherzer,
white}@uwf.edu, {gng3,CTerry}@students.uwf.edu

Arthur Baskin
Intelligent Information Technologies, Indianapolis, Indiana, U.S.A.
e-mail: abaskin@intelligent-it.com

P.-J. Benghozi, D. Krob, & F. Rowe (Eds.): *Digital Enterp. Des. & Manage. 2013*, AISC 205, pp. 45–56.
DOI: 10.1007/978-3-642-37317-6_5

1. As companies form partnerships and governments strive to integrate the work of different departments (Janssen et al. 2011), business processes become ever more interconnected even across organizational boundaries.
2. Each step in such processes depends ever more heavily upon software support, usually involving pre-existing information systems developed using diverse standards and technologies.

It is clearly unrealistic to design a new technologically harmonized information system to meet each emerging need. So the only practical solution is to find ways to allow existing, technologically diverse systems to interoperate. Interoperation has been described as having two or more independent systems operate in a coordinated and meaningful fashion such that processes are effectively merged or information is effectively shared (Scholl and Klischewski 2007).

While there have been many attempts to achieve interoperability, the approach that seems to have the most followers at the moment is Services Oriented Architecture (SOA), usually implemented following the Web Services (WS) standards (Josuttis 2007 Chapter 16). SOA is not regarded as a specific architecture but rather as a general way of structuring software. Terminology varies but typically *composite applications* are constructed by *orchestrating* loosely coupled *services* running on different nodes and communicating via message passing. Ideally each service represents a discrete business function that can act as a reusable component across multiple applications. Often an infrastructure layer, sometimes called an *Enterprise Service Bus* (ESB), mediates service interactions providing functions such as message routing, reliable messaging and data transformations.

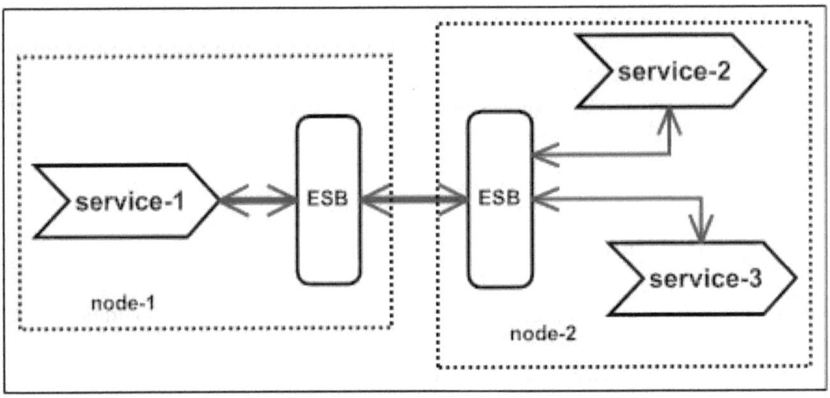

Fig. 1 Structure of a SOA Composite Application

SOA/WS has the potential to be a transformational technology with widespread impact on the way humans live. Earlier transformational technologies, such as printed books or steam-powered transportation, broke down barriers that prevented connections; in the one case connections of ideas from human to human, in the second connections of goods between producers and consumers. As connections become feasible, interactions multiply and new human opportunities emerge.

In theory, SOA/WS may become transformational if it can enable connections between data and processing, both within and across organizational boundaries. However experience with SOA/WS has been somewhat mixed. Both organizational and technical obstacles have sometimes made it difficult to achieve the promised benefits of this approach (Luthria and Rabhi 2012). In this paper we will discuss one such obstacle that is one of the oldest problems in Software Engineering, that is, the cost and delays of software evolution.

A distinguishing characteristic of interoperable composite applications is that ownership and control over services is also often distributed. Whether the different services are contained within one large organization (such as the many agencies within a government) or whether some are completely outsourced (as with commercial cloud services), the IT manager must deal with components that are not under his control. Evolution may then be driven by external changes on an inflexible schedule.

> For example one of the authors of this paper is the developer of a composite application that gathers price data from several sources, including the Amazon Product SOAP API to prepare a pricing guide for a retail business. Recently Amazon has announced that, in a few months' time, this interface will be restricted to Amazon affiliates only. The developer must now evaluate where and how he is using that particular service and develop an alternative or work-around.

If IT managers are to commit to interoperable systems, they must first have confidence that they can respond to such external changes with agile software evolution. In this paper we will discuss some of the technical challenges in the evolution of interoperable applications and describe our ongoing research into support tools based on enhanced search technology for SOA/WS.

2 Software Evolution and Interoperability

The term *maintenance* has traditionally been applied to all changes that are performed upon a software system after its deployment. However since most changes are not bug fixes, some have proposed that we should instead call this process *software evolution* and this may indeed be a better term to describe a system's overall history of change. However "evolve" is a passive verb so it misses the fact that software change requires work by talented and costly professionals. Perhaps a compromise would be to say that the software *evolves* because of the work Software Engineers perform to *maintain* it and so we will use both terms in this paper.

The defining characteristic of software maintenance/evolution as opposed to development is that any proposed change needs to take into account a large base of existing software. The change is thus highly constrained; a Software Engineer cannot just look for the best way to implement a requirement but rather must look for the best way given the existing design and code base. Changes are costly and hard to plan because of uncertainties in the time required to understand the existing system, plan a change, identify impacts of the change and test that all impacts have been correctly handled. It is not uncommon for even experienced Software Engineers to grossly underestimate the ripple effects of a software change.

Several authors have pointed out that SOA composite applications may present a particular challenge for program comprehension and thus for software maintenance (Gold et al. 2004, Canfora and Di Penta 2007, Lewis and Smith 2008, Kontogiannis 2008). Most of the factors mentioned stem from or are exacerbated by the distributed ownership of the different services:

1. Services making up a composite application may use diverse technologies: in their operating system, in their programming languages, and in their messaging layer. This diversity complicates invocation across system boundaries. The Web Services standards attempt to mitigate these difficulties by prescribing the use of Web Services Description Language (WSDL) to specify service interfaces with XML Schema Descriptions (XSDs) used to describe data passed in messages. Given a WSDL and its XSDs, tools can construct proxies and stubs allowing code using one technology to invoke code using another. However the tools are not currently problem free (McGregor et al. 2012) and in any case the WSDL really only covers the mechanics for message exchange. It does not provide any information about how service invocations need to be sequenced, the state changes produced by a service invocation, and other semantic factors that may be essential for program comprehension.
2. The services may employ different semantics, especially subtly conflicting meanings of data items, which makes it problematic for one system to use data provided by another. For example, Gold and Bennett mention that the term "child" has numerous different meanings in different organizations in the United Kingdom health care domain (Gold and Bennett 2004).
3. Composite applications may face complex and changing security requirements: some operations may be restricted, data may need to be confidential and all actors in the system should not have the same access. This means that maintenance changes need to be looked at carefully not only from the point of view of their functionality, but also as to their application-wide effects on security.
4. For commercial reasons the owners of a service may not choose to make available complete documentation of design, defects encountered, change history, and so on. Software Engineers in the consumer organization may thus have greater difficulty understanding changes in service behavior.

3 Program Comprehension for SOA Evolution

Thus while interoperability is a business necessity, it brings with it increased complexity and some level of loss of control. The challenge for SOA evolution then is to perform needed changes quickly and correctly despite these new factors. The key roadblock is the same as for evolution of traditional systems: a Software Engineer has to understand a program in order to change it effectively and safely.

An obvious goal then is to reduce the costs of SOA program comprehension, specifically by providing tools and techniques that can help Software Engineers navigate and understand the artifacts making up a composite application. This goes beyond studying code to analyze artifacts such as:

1. Descriptions of both external and internal services; WSDLs and XSDs together with documentation in any format provided by the owner.
2. Deployment configuration files such as the *web.config* of ASP.NET, the *web.xml* of J2EE and numerous others that determine how services are accessed, how the execution environment is configured, what security restrictions are set, etc.
3. Middleware configuration, such as configuration files for a particular ESB, application server-specific configuration such as *sun-web.xml* (for GlassFish) etc.

Clearly one challenge is the “open world” nature of SOA (Van den Heuvel et al. 2009). Many of the artifacts mentioned are specific to a particular technology or vendor. Yet technologies, vendors, and the specific set of partners in a composite are likely to change greatly over the application's life cycle. We cannot predict an exact static mix of technologies, programming languages, and documentation formats that will remain valid for the life of any particular composite application. Indeed, it is likely that maintenance of each mature SOA composite application will involve a somewhat different and changing combination of artifacts.

This means that support tools for SOA evolution must also be able to function in this open world. Somehow we need to create support tools that will have the flexibility to adapt just as the application itself adapts.

There has been relatively little published work on tools to support program comprehension for SOA. Most of that work has concentrated on dynamic analysis tools to analyze patterns of execution from a running system instead of studying the artifacts that describe it. For example a group at IBM has developed a *Web Services Navigator* visualization tool that captures event logs from a running system and analyzes the resulting data to identify logic and performance problems (De Pauw et al. 2005). Another dynamic approach recovers a sequence diagram showing how a particular user feature executes. It does this by comparing inter-process messages collected when the feature is running with background messages collected when the system is performing other tasks (Coffey et al. 2010). For testing of an external service, another proposal is to start with a model of the expected sequence of interactions and generate test cases which are used to probe the running service and confirm or reject the model (Halle et al. 2010).

Dynamic analysis approaches to program comprehension have substantial advantages, especially in visualizing the complex, dynamically-changing interaction patterns of a SOA composite application. They do, however, require preparing tests and setting up a running copy of the application. The environment must support collecting and correlating traces or logs from multiple nodes. There is thus a considerable amount of set-up work required, which may need to be repeated if the partners or technology of the application change.

4 Basic Search for SOA Evolution

Our research group has been focusing on search technology as a possibly simpler foundation for tools for SOA's open world. Text search has the advantage of being a well-established and well-known paradigm for gathering knowledge, as is evidenced by the overwhelming adoption of search services such as Google™ and Bing™. Search allows the efficient collection of information across a wide range of sources and document types, though the cost of this generality is that almost all the work of comprehension is put upon the user.

We have been experimenting for some time with a specialized search system for SOA evolution called *SOAMiner*. Our tool indexes a large collection of artifacts related to a particular SOA composite application and allows queries to locate information to support a particular maintenance task.

The original motivation for SOAMiner came from studying a small SOA composite application provided as a tutorial for a well-known open-source Integrated Development Environment (IDE). While intended to be a simple example, the whole application once deployed consisted of no less than 129 files distributed across 49 directories, not counting files actually deployed to the server! The most important artifacts were WSDL interface specification files backed up by XSDs for the industry-standard data types that were passed in the messages. The services were orchestrated by code in Business Process Execution Language (BPEL).

We found that it was very difficult to navigate the many interconnections between and within these artifacts and thus understand the overall structure of the composite application. Using the IDE helped somewhat, but relying on it would mean that any company owning the application would be dependent on that particular tool vendor. (As it happened, within a year and after a corporate take-over, the BPEL features of this particular IDE were no longer available.)

We noted that all three kinds of artifacts (WSDL, XSD, and BPEL) have XML structure and most of the information about interconnections was contained in attributes within the XML tags. Thus the first prototype of SOAMiner focused on extracting and searching text from such tags and was applicable to any file with XML structure[1]. We performed some initial studies using small datasets to let students evaluate the usability of the tool and larger ones to confirm acceptable performance (White et al. 2011).

5 Knowledge-Enhanced Search for SOA Evolution

Our initial SOAMiner studies showed that the search approach was very powerful in locating information in the large corpus of artifacts of a SOA composite application. Yet it left the user with the often difficult task of understanding the results from each search query. The search simply displayed the XML tags that matched

[1] An online demonstration of this initial prototype may be viewed at `http://soademo.cs.uwf.edu/SOASearch/`

the query so the user had to supply the mental context of each match and often had to make multiple searches to trace through the relationships.

As an illustration, consider trying to understand service relationships in a simple system such as WebAutoParts.com, a SOA composite application that we have used in some of our case studies (Figure 2). WebAutoParts.com is a hypothetical online automobile parts supplier (Wilde et al. 2012).

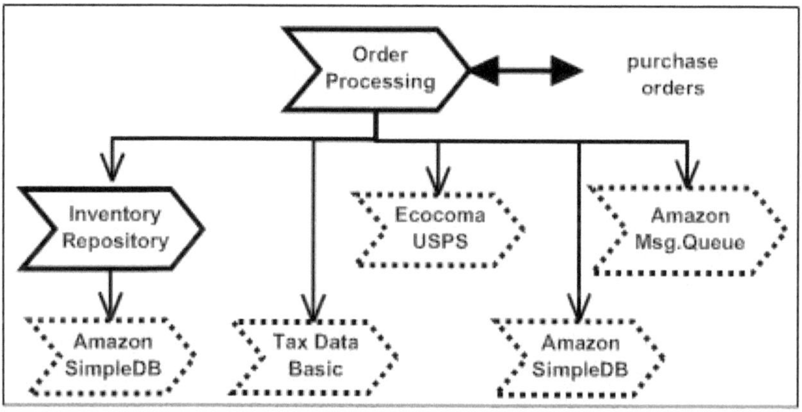

Fig. 2 WebAutoParts Internal(solid) and External(dotted) Services

The WebAutoParts order processing workflow of Figure 2 has two stubbed in-house services in BPEL (Order Processing and Inventory Repository) and four external services represented by WSDLs and XSDs from three well-known vendors:

1. Amazon Web Services - Amazon Simple DB (database) and Message Queue (message queuing)
2. StrikeIron.com - Tax Data Basic (sales tax rates)
3. Ecocoma - USPS (shipping costs)

In this work flow, a purchase order is received, inventory availability is checked using the Inventory Repository service, American state sales tax is added depending on customer location, and USPS shipping costs are computed. The purchase order is then stored using Simple DB and a message is sent via the Message Queue service to trigger order fulfillment at the warehouse.

Suppose a Software Engineer unfamiliar with this application is trying to implement a change to the database design and needs to know what data is passed when Order Processing checks inventory levels. If he has extensive BPEL/Web Services experience he might figure this out using a series of searches:

1. Search the Order Processing BPEL file to find the <invoke> tag that is checking inventory. That provides him a partnerLink. Then search the partnerLinks to get the partnerLinkType which turns out to be IRepositoryLinkType.

2. However the BPEL provides no indication of which service implements this link type, so the Software Engineer now searches all the WSDL documents for that link type. He will find it in InventoryRepositoryArtifacts.wsdl with a pointer to the WSDL portType for the service. The portType in turn gives the <operation> tag and its input and output message names. A further search on the message name reveals that the message contains an element called inventoryQuery.
3. However inventoryQuery is not defined within the WSDL so the Software Engineer now has to search XSDs to eventually locate the definition of inventoryQuery, determine its type, and from its type finally conclude what data fields are being passed.

This is, of course, just one example of the many relationships a Software Engineer may need to navigate to be able to understand a SOA composite application. The first prototype of SOAMiner greatly facilitates such searches, but not the process of establishing the relationships.

Clearly what is needed is some form of reverse engineering or abstraction to aid in comprehension. Where possible we would like to be able to search the artifacts making up a SOA application and then return abstractions that would quickly provide meaningful information to a maintainer. This raises two questions:

1. What are the important abstractions for SOA maintenance/evolution?
2. How can we provide abstractions while living with the changes of the SOA open world?

To identify important abstractions we conducted two case studies using the SOAMiner prototype. Both studies were informal; a small number of participants were asked to answer questions about a SOA system using the prototype while "thinking out loud" about their actions. They were observed while performing their task and then debriefed afterwards to capture their impressions and suggestions. The questions were chosen based on the kinds of search that Software Engineers have been found to use while developing and maintaining pre-SOA software systems (Sim et al. 1998).

The complete design and results of the case studies have been reported elsewhere (Reichherzer et al. 2011, White et al. 2012). However three abstractions stood out among the suggestions from study participants:

- **A compact representation of a service:** A WSDL file with its associated XSDs provides a representation of a service interface that is very difficult for humans to handle. In the great majority of cases the service could be displayed as a simple tree showing the service, its operations, the input/output messages for each operation, and the data types for each message.
- **Compact data type summaries:** XML Schema Descriptions (XSDs) are used in WSDLs to describe the data passed in messages; the XSD tags may be incorporated directly into the WSDL or else imported from separate files. In both cases, the data description may be complex and dispersed with multiple levels

of type and element descriptions which reference each other. For most cases a tree representation or an E-R diagram could be constructed that would be far easier for Software Engineers to grasp.
- **BPEL invoke relationships:** As shown by the WebAutoParts example given earlier, it may take a long sequence of steps to trace out what services are actually being invoked from a given BPEL file. It would be possible to automate this trace to draw a tree representation of the invoke relationships and provide a picture similar to Figure 2.

Our second question was how to compute these and other abstractions given the open and evolving nature of SOA composite applications. We needed some flexible and extensible way of defining abstractions over XML artifacts.

We are currently experimenting with an add-on to SOAMiner that uses a rule-based system to compute abstractions from a forest of XML-structured documents. Each abstraction type is specified as a set of rules that describe how a tree-representation of a specific abstraction may be derived from the original XML. For initial tests we have implemented rules for the three abstractions identified in the case studies. It should be easy to add or remove rules to allow the search tool to adapt to different technologies.

6 Current Directions for SOA Evolution Support

Our group is currently researching two additional approaches to enhance support for SOA evolution.

1. Searching code and documentation
2. Incorporating ontological information to improve system comprehension

Our initial prototype of SOAMiner focused on finding effective ways to search files with XML structure since they are most characteristic of SOA. WSDLs, XSDs, BPEL and many deployment and middleware configuration files all have an XML structure. A key decision was choosing the right granularity for indexing and results. SOAMiner treats each tag in the corpus as a separate entity, since it would not be very useful to respond to a query with an entire WSDL which may contain hundreds of tags.

An obvious way to increase the effectiveness of the search would be to also search source code and documentation. Here again the key question will be to establish the granularity for search. Documents may have no structure at all, or a very loose HTML or Microsoft Word structure. Code on the other hand is highly structured, but there are many different structures depending on the syntax of the specific programming language. Unfortunately SOA code exists in numerous languages and versions and it is impractical, at least for a research project, to provide a tool which will parse them all. As well, a tool that depends too strongly on specific syntax would not seem appropriate for the SOA open world since it may rapidly become obsolete as languages change.

Our research is experimenting with ways of deconstructing code and documentation into meaningful fragments that balance the need for tool generality against the increased power that can be obtained by taking advantage of input structure.

A second research direction is to provide ontological support for SOA program comprehension. An ontology represents the set of concepts in a particular domain, together with the relationships between those concepts. Our group is working to prepare a set of ontologies that represent those concepts that may be important for a Software Engineer performing maintenance on a specific SOA composite application.

The Open Group has released an ontology for Service-Oriented Architecture which is very useful for describing the overall structure of a composite application, its actors, its services, and its data (Open Group 2010). However since it is largely technology-neutral it does not go into many of the implementation details that a maintainer would need to deal with. Also, for any particular SOA composite it would be useful to have a domain ontology that describes the relationships in the real world within which the software is operating.

Ontological descriptions could be useful by themselves, for example to provide a shared vocabulary for human discussion of a system. They could also be useful to enhance the usefulness of search tools such as SOAMiner and its rules-based extension. Search could be enhanced by providing synonyms for search queries, by creating cluster abstractions of related software elements, and by prioritizing the display of search results based on semantic information.

7 Conclusions

In this paper we have argued that SOA has the potential to be a truly transformational technology as it opens up new opportunities for interoperability between software systems. But by their nature interoperable systems imply an environment of distributed ownership and control. Managers will be reluctant to trust such an environment unless they can be confident in their ability to respond to external changes with agile software evolution, but agile evolution requires rapid program comprehension of complex and heterogeneous systems in an open world with changing partners and changing technologies.

Our group's approach has been to focus on flexible, enhanceable search-based tools for program comprehension. The tool objective should be to be smart where possible, but useful everywhere, so that tool performance degrades gracefully as SOA technologies change. Thus for SOAMiner, the eventual goal would be to:

1. Provide abstraction-enhanced and ontology-enhanced search where it can
2. Provide useful text-based search everywhere else
3. Progressively allow more searches to be moved into the first category.

This is only one possible research direction out of many. Model Driven Architecture (MDA), for example, is an approach to SOA development that could have significant value for maintenance. With MDA a composite application is defined by models; code is either generated automatically or else the model may be directly interpreted at runtime (Salhofer and Stadlhofer 2012). The program comprehension burden may be reduced if only models need to be maintained and if they are substantially easier to understand than code. Of course the challenge may be to compatibilize the MDA models across a system with distributed ownership.

Ideally, definition of practical tools to support SOA evolution should be a collaboration between industry and researchers. The range of questions that SOA maintainers will face is still far from clear and an industry-university dialog could be most useful. We have been working with industrial contacts in the Security and Software Engineering Research Center (S2ERC) but a broadening of the conversation could help target research on the most important practical obstacles to agile change. Better tools and methods for SOA evolution could then be a major enabler for the interoperable systems of the future.

Acknowledgments. Work described in this paper was partially supported by the University of West Florida Foundation under the Nystul Eminent Scholar Endowment and by the Blue Cross Blue Shield Association and by Intelligent Information Technologies, both industrial affiliates of the Security and Software Engineering Research Center (www.serc.net).

References

Canfora, G., Di Penta, M.: New Frontiers of Reverse Engineering. In: Proc. Future of Software Engineering, pp. 326–341 (2007), doi:10.1109/FOSE.2007.15

Coffey, J., White, L., Wilde, N., Simmons, S.: Locating Software Features in a SOA Composite Application. In: Proc. 2010 Eighth IEEE European Conference on Web Services, ECOWS 2010, pp. 99–106 (2010), doi:10.1109/ECOWS.2010.28

De Pauw, W., Lei, M., Pring, E., Villard, L., Arnold, M., Morar, J.F.: Web Services Navigator: Visualizing the execution of Web Services. IBM Systems Journal 44(4), 821–845 (2005), doi:10.1147/sj.444.0821

Gold, N., Bennett, K.: Program Comprehension for Web Services. In: International Conference on Program Comprehension. IEEE Computer Society (2004), doi:10.1109/wpc.2004.1311057

Gold, N., Knight, C., Mohan, A., Munro, M.: Understanding Service-Oriented Software. IEEE Software 21(2), 71–77 (2004), doi:10.1109/MS.2004.1270766

Halle, S., Bultan, T., Hughes, G., Alkhalaf, M., Villemaire, R.: Runtime Verification of Web Service Interface Contracts. IEEE Computer 43(3), 59–66 (2010), doi:10.1109/mc.2010.76

Janssen, M., Charalabidis, Y., Kuk, G., Cresswell, T.: E-government Interoperability, Infrastructure and Architecture: State-of-the-art and Challenges. Journal of Theoretical and Applied Electronic Commerce Research 6(1), i–vii (2011), doi:10.4067/S0718-1876 2011000100001

Josuttis, N.M.: SOA in practice: The art of distributed system design. O'Reilly (2007) ISBN 0-596-52955-4

Kontogiannis, K.: Challenges and opportunities related to the design, deployment and operation of Web Services. In: Proc. Frontiers of Software Maintenance, pp. 11–20 (2008), doi:10.1109/FOSM.2008.4659244

Lewis, G.A., Smith, D.B.: Service-Oriented Architecture and its implications for software maintenance and evolution. In: Proc. Frontiers of Software Maintenance, pp. 1–10 (2008), doi:10.1109/FOSM.2008.4659243

Luthria, H., Rabhi, F.A.: Service-Oriented Architectures: Myth or Reality? IEEE Software 29(4), 46–52 (2012)

McGregor, S., Russ, T., Wilde, N., Gabes, J.P., Hutchinson, W., Duhon, D., Raza, A.: Experiences Implementing Interoperable SOA in a Security-Conscious Environment. S2ERC-TR-307, Security and Software Engineering Research Center (S2ERC) (June 6, 2012), http://www.serc.net; also available at http://www.cs.uwf.edu/~wilde/publications/TecRpt307/ (accessed July 2012)

Open Group, Service-Oriented Architecture Ontology (2010) ISBN 1931624887, https://collaboration.opengroup.org/projects/soa-ontology/?gpid=483

Reichherzer, T., El-Sheikh, E., Wilde, N., White, L., Coffey, J., Simmons, S.: Towards intelligent search support for web services evolution: identifying the right abstractions. In: 13th IEEE International Symposium on Web Systems Evolution (WSE 2011), September 30, pp. 53–58 (2011), doi:10.1109/WSE.2011.6081819

Salhofer, P., Stadlhofer, B.: Semantic MDA for E-Government Service Development. In: 45th Hawaii International Conference on System Sciences, pp. 2189–2198 (2012), doi:10.1109/HICSS.2012.524

Scholl, H.J., Klischewski, R.: E-Government Integration and Interoperability: Framing the Research Agenda. International Journal of Public Administration 30(8-9), 889–920 (2007), doi:10.1080/01900690701402668

Sim, S.E., Clarke, C.L.A., Holt, R.C.: Archetypal source code searches: a survey of software developers and maintainers. In: Proc. 6th International Workshop on Program Comprehension, IWPC 1998, pp. 180–187 (1998), doi:10.1109/WPC.1998.693351

van den Heuvel, W.J., Zimmermann, O., Leymann, F., Lago, P., Schieferdecker, I., Zdun, U., Avgeriou, P.: Software service engineering: Tenets and challenges. In: PESOS 2009, May 18-19, pp. 26–33 (2009), doi:10.1109/PESOS.2009.5068816

White, L.J., Reichherzer, T., Coffey, J., Wilde, N., Simmons, S.: Maintenance of service oriented architecture composite applications: static and dynamic support. J. Softw. Maint. Evol.: Res. Pract. (2011), doi:10.1002/smr.568

White, L., Wilde, N., Reichherzer, T., El-Sheikh, E., Goehring, G., Baskin, A., Hartmann, B., Manea, M.: Understanding Interoperable Systems: Challenges for the Maintenance of SOA Applications. In: 45th Hawaii International Conference on System Sciences, pp. 2199–2206 (2012), doi:10.1109/HICSS.2012.614

Wilde, N., Coffey, J., Reichherzer, T., White, L.: Open SOALab: Case Study Artifacts for SOA Research and Education. In: Principles of Engineering Service-Oriented Systems, PESOS 2012, Zurich, Switzerland, June 4, pp. 59–60 (2012), doi:10.1109/PESOS.2012.6225941

Enterprise Architecture: Beyond Business and IT Alignment

Marcel Lee

Abstract. IT has always embodied both a huge opportunity and a misunderstood asset in most enterprises. Today, pressure has never been so hard on IT executives to reduce cost and complexity while bringing value. Enterprise architecture was born in an attempt to address this very challenge. But while serving its original purpose, it usually focuses on business and IT, overlooking other aspects potentially required for envisioning enterprise transformation. The latter is becoming more and more compulsory and IT has never been so pervasive in today's enterprises. In this context, approaching transformation through a business/IT duality as in most enterprise architecture methodologies proves to be insufficient. Pervasive IT infers the consideration of other aspects, all intermingled one another. One possible way to deal with this reality is to model the enterprise as a graph of aspects, beyond business and IT. By putting weight on the relationships between those aspects, one can delineate a Minimal Spanning Tree that would constitute a pragmatic yet complete frame of analysis. Applied to an imaginary business case, the aforementioned approach proves to be relevant for enterprise analysis in a holistic yet pragmatic way. It can also be integrated in existing frameworks, either as an extension, or as an overarching frame for a hypothetic *enterprise transformation practice*. Yet, many additional works need to be achieved before envisaging such a practice in the future.

1 Searching for Alignment

1.1 Original Purpose of Enterprise Architecture

IT has always been perceived as both a huge opportunity and an immense burden by organisations. When cleverly used, it can greatly improve performance and

---

Marcel Lee
Pramana
99 avenue de la République
75011 Paris, France
e-mail: marcel.lee@pramana.fr

P.-J. Benghozi, D. Krob, & F. Rowe (Eds.): *Digital Enterp. Des. & Manage. 2013*, AISC 205, pp. 57–66.
DOI: 10.1007/978-3-642-37317-6_6

foster innovation. In many ways, it also embodies an intangible and mostly misunderstood asset whose value is always to be proven, complexity to be tamed and cost to be put under control.

In the early days, organisations have rushed into IT to take advantage of its potential: process automation, digital desktop, etc. All these advances have been devised and implemented through most of the time an unplanned and rather opportunistic way. This led to a rather chaotic IT landscape without a solid foundation to enable reuse and homogenisation. Over time, the cost and complexity of IT systems have exponentially increased while the additional business value to be derived from those systems has constantly decreased. Organisations have found themselves facing the following challenge: how can I derive more value from IT and at the same time reduce cost and complexity? This question has been addressed by several areas of expertise, among which, enterprise architecture.

1.2 Serving the Business and Aligning IT

If one cannot find a unique commonly agreed definition for enterprise architecture, practitioners usually do agree on the benefits it can bring to an organisation: it facilitates the reduction of complexity through better interoperability and promotes cost reduction, through better return on investments and flexibility of business capabilities. Practitioners also do usually agree that enterprise architecture should "enable the process of translating business vision and strategy into effective enterprise change" [1]. The exact modalities of how to proceed differ from one methodology to another but most of them are usually applied with the following commonality: a *business/IT duality*, with a strong emphasis on the necessity for *IT to be aligned with the business*. Let's examine in detail the practices mostly in use in organisations today to confirm this statement.

Enterprise architecture is usually applied in a top-down approach. It operates with a strong support from the top management in an environment where the vision statement is clear. In this configuration, all the analysis is conducted so that the target state of the organisation is in line with the business vision. This is usually conducted using the "separation of concerns" [2] principle, the concerns in study differing according to the methodology in effect.

For example, in TOGAF (The Open Group Architecture Framework [3]), the concerns in question are the Business Architecture, Information Systems and the Technology Architecture. There is a clear business/IT duality in the way architecture is envisaged and IT is supposed to support the target state of the Business Architecture, which should itself be directly aligned with the strategic objectives.

In the Zachman framework [4], even though the separation between business and IT is less formal, the first two audience perspectives (executive and business management) represent the business aspect whereas the three following audience perspectives (architect, engineer, and technician) represent the IT aspect. The alignment notion is clearly represented as a link to be made alongside the audience perspectives. Like in TOGAF, little or no attention is turned to other concerns than business and IT.

This business/IT duality is even more striking when adopting a "city planning" layered approach to enterprise architecture. It organises enterprise into a layered hierarchy, from the business layer to the technical layer. This methodology, while aligning IT with the business by essence, gives however few or no room for the study of other concerns than business and IT.

Now that we have examined the usual practices in effect in enterprise architecture, it becomes apparent that there is usually a strong common emphasis on business/IT alignment which permits to prioritize initiatives according to the value they can bring. Moreover, there is systematically a strong emphasis on reuse and the building of a foundation architecture. Therefore, enterprise architecture as practised today is serving well its main original purpose: reducing cost and complexity while deriving more value from IT.

This business/IT duality actually reflects quite well the usual state of mind of C-level executives regarding IT, namely its contribution to the business vision and strategy in a context where IT is mostly seen as a utility or as a service provider to the business. But only very few methodologies approach the enterprise in a context of overall transformation [5], embracing concerns out of the business and IT scope, like culture, skills or behaviour. Only in the Praxeme framework can we see a more global approach, as enterprise architecture is defined as the place "where the effort is made to consider all the aspects together" [6]. But are all these other concerns of primary importance when dealing with enterprise change? And if so, how can we cope with them?

2 Intertwining Aspects

2.1 Ubiquitous IT: A Reality

Technology has never been so present in our everyday's life. With social network, mobility, and new media all in play, our relation with technology has been reshaped for good. This trend is also increasingly spreading in organisations. While confined in the early days in a limited number of functions, technology is now omnipresent: we use applications to develop customer relationship, we use mobile devices and digital signature to deliver parcels, and we use social networks to develop new forms of marketing. In a sense, IT has become ubiquitous in most organisations.

2.2 Transforming the Enterprise Beyond Business and IT

But ubiquitous IT is not all. Organisations have never faced such a shifting and unstable environment. With the advent of new regulatory policies and the ever harder pressure of the competition, organisations are constrained to transform constantly.

In this context of ubiquitous IT and compulsory transformation, is enterprise architecture as practised today relevant for an effective translation of business vision and strategy into effective enterprise change? As discussed in the first chapter of the present paper, enterprise architecture usually addresses transformation by approaching business and IT, the main concern being the alignment of IT with the business. This approach is suitable in a vision where the business is the most important driving force of the enterprise and where IT represents an asset whose main function is to serve the business. But when IT is ubiquitous and when interactions exist between all aspects, this simple approach may be inadequate for practical purposes. Indeed, how many times have we witnessed a transformation fail because some critical aspects like culture or behaviour were overlooked? And yet, in those cases, enterprise architecture was quite practised as defined by the state of the art methodology.

2.3 Embracing Other Subjects

Let's now examine the disciplines involved with transformation in a context where *IT is ubiquitous* in the organisation. Most often, the organisation proceeds through an association of disciplines such as strategy, business analysis, innovation, enterprise architecture, change management, and talent management. People all work together to enable the transition into the desired state of the organisation and to find the right interaction that delivers added value.

Needless to say, there are other disciplines involved in the transformation process and a strong "organic" interaction between them. These disciplines imply quite a number of aspects beyond business and IT as well as interactions in both directions. For example, the organisational behaviour influences the way people interact within business processes which in turn influence the target state of the IT landscape. Conversely, the latter influences skills and competencies. This illustration calls for two observations:

a) All aspects are intertwined and to be considered when transforming an organisation, beyond the separation wall of business and IT
b) Enterprise architecture is not the only discipline at stake when transforming the organisation

Given this, can enterprise architecture take up the challenge of mingling different aspects of the organisation beyond business and IT and act as more than a catalyst for transformation?

2.4 Enterprise as a Network of Aspects: A Holistic Approach

In order to answer the latter question, let us formalise architecture in the context of an enterprise. Architecture can be defined as the "structure of components, their **inter-relationships**, and the principles and guidelines governing their design and evolution over time" [7]. An enterprise can be defined as an "organisation engaged in the trade of goods, services, or both to consumers" [8] and viewed as a complex system of socio-technical aspects including people, culture, information, technology, business operations, and intellectual property. Enterprise architecture is usually focused on the business and IT aspects of the organisation. This scope is surrounded by a context, which includes all other aspects. Enterprise architecture is normally responsible for the articulation between the scope and the context as described in the following figure, while making the necessary alignment between business and IT.

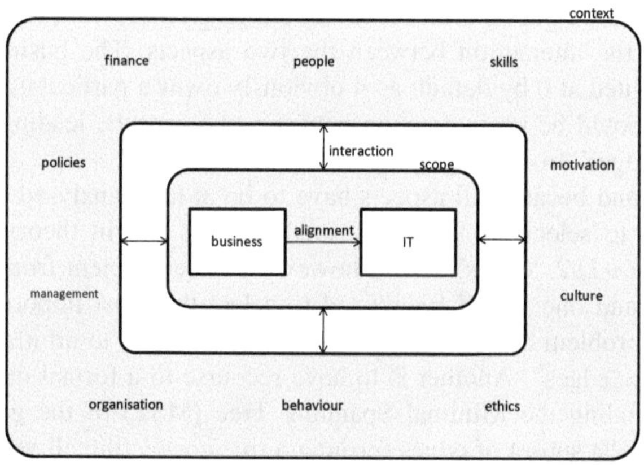

Fig. 1 Usual scope and context of enterprise architecture

While simple in its conception, this way of separating concerns may be out of phase with reality. We indeed observed that in a context of transformation, all aspects are to be considered and are linked with one another. This would imply an overlay of the scope and context as well as links between all aspects as follows.

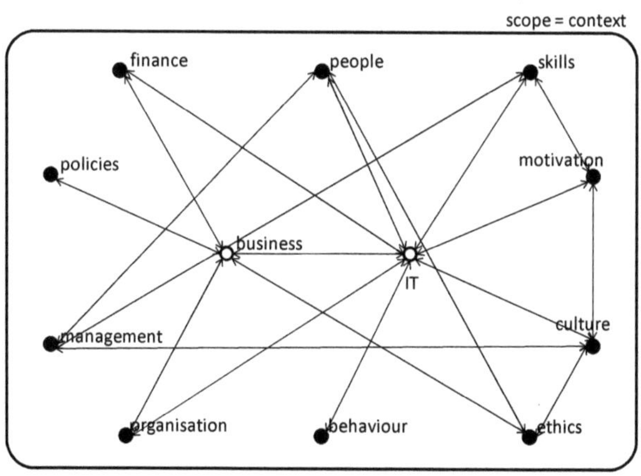

Fig. 2 The enterprise: a network of intertwined aspects

In graph theory, the above representation is an undirected and unweighted graph. Each aspect in consideration can be represented by a "vertex", and each link between two aspects can be represented by an "edge". Using a **holistic approach**, approaching enterprise architecture would theoretically imply studying *n* "vertices" and *n(n-1)/2* unique "edges", in two temporal states (baseline and target).

One matter yet to consider is that depending on the case, interactions between aspects are in reality not equivalent one another. For example, while in most enterprises there is a strong mutual influence between "business" and "behaviour", there can be very few interactions between "IT" and "ethics". This can be modelled by putting a weight on each "edge" of the graph inversely proportional to the importance of the interaction between the two aspects. The business/IT "edge" would be weighted at 0 by default as it obviously owns a particular position. Conversely, there could be no interaction between two aspects, leading us to simply delete those "edges" from the graph.

Given this, and because all aspects have to be at least analysed separately, we are compelled to select all "vertices" in the graph and in theory, all "edges". Studying the *n(n-1)/2* "edges" would however not be efficient from an executive point of view and one would be enticed to select the most important remaining "edges". This problem can be solved in two ways. One is to arbitrarily select the most important "edges". Another is to have recourse to a formal method, such as for example finding the Minimal Spanning Tree (MST) of the given graph. A spanning tree is "a subset of edges forming a tree connecting all vertices" and the MST is "the smallest connected graph in terms of edge weight" [9]. Among the possible algorithms to find the MST, stand the Prim's and the Kruskal's algorithm [9]. We consider the Kruskal's algorithm since it will automatically by principle select the business / IT "edge" whose weight is 0 by default. The following figure illustrates this algorithm on a graph representing the interconnected aspects of the organisation in one given temporal state.

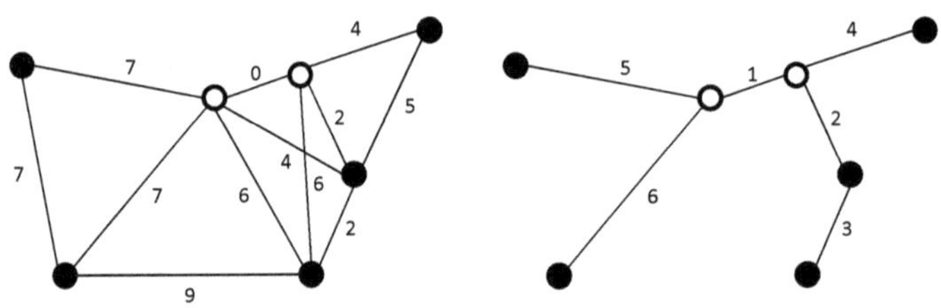

Fig. 3 Illustration of the Kruskal's algorithm. The numbers in the original graph represent weights. The numbers in the MST represent the order of insertion.

This example illustrates that it may be conceivable to reduce the number of interactions to study while at the same time address the most important ones using a formal graph algorithm. Consequently, one could envisage tackling the analysis part of enterprise architecture in its entirety and not uniquely on business and IT aspects. This holistic approach may have the benefit of broadening the scope of enterprise architecture and revive its purpose in today's environment. It could thereby have the potential to become one day the practice to consider when dealing with enterprise transformation as a whole.

In the next chapter, we make an attempt to use this approach on an imaginary business case before giving hints for new perspectives.

3 Bringing It into Reality

3.1 Framing the Analysis

In order to illustrate the approach described above, we now imagine a business case with an enterprise named Buy-A-Lot, a major player in the electronics retail sector. In 2012, Bob Casey, the CEO of the company, in response to a harsh competing environment, decides to launch the "Back on Track" program based on three major strategic orientations: cooperation, interoperability and agility.

In order to plan the transformation, we define the target state of the company using first a traditional business/IT dual framework, leading to the following high-level vision statement.

Table 1 High-level vision of the target state of Buy-A-Lot on Business and IT

Aspect	Description in the target state
Business	Highly tactical customer services Harmonised and end-to-end processes Clearly defined business services
IT	Agile development Portals and Intranet Mutualisation of technical resources Usage of standards SOA (Service Oriented Architecture)

This first analysis could be satisfying as such and lead to further analysis and modelling. Bob Casey is however warned about common SOA pitfalls, especially those dealing with changes in business habits, including the reduction of overlapping in roles and responsibilities. On that account, he wishes to build analysis on the following aspects as well: behaviour, organisation and skills, making a number of 10 interactions to study. By adopting the MST approach, he hopes that this number and consequently the amount of work to provide could be reduced.

The following table is the weighted adjacency matrix of the aforementioned aspects in the target state of Buy-A-Lot. Weights are attributed based on the assumption that SOA leads to fundamental changes in behaviour and skills while harmonised and end-to-end processes infer changes in behaviour and organisation.

Table 2 Weighted adjacency matrix of the aspects in the target state of Buy-A-Lot

	Business	IT	Behaviour	Organisation	Skills
Business	-	∞	12	10	5
IT	∞	-	15	4	8
Behaviour	12	15	-	3	2
Organisation	10	4	3	-	4
Skills	5	8	2	4	-

Using Kruskal's algorithm, we can delineate the MST of the graph that leads us to reduce the number of interactions to study by 60%. The outcome of the discussion on the four remaining interactions is as follows:

Table 3 Discussion on the main interactions between aspects of Buy-A-Lot

Edge	Discussion on the alignment
Business-IT	- Already covered in the first analysis
IT-Behaviour	- SOA calls for responsibility and transparency - People should collaborate in a service oriented fashion between - departments (no more request isolated from predefined services) - Mutualisation of resources and standards mean more strict rules and potential tensions - Agile development infers perimeter and timing strictness
Business-Organisation	- Clearly defined business services and end-to-end processes imply no redundancy in the organisation
IT-Skills	- New skills are needed with agile development, SOA and collaboration through portals and intranet

This quick analysis makes it possible to refine the description of the aspects and draw a more consistent vision of the target state of the enterprise.

Table 4 Refined high-level vision of the target state of Buy-A-Lot

Aspect	Description in the target state
Business	Highly tactical customer services Harmonised and end-to-end processes Clearly defined business services
IT	Agile development Portals and Intranet Mutualisation of technical resources Usage of standards SOA (Service Oriented Architecture)
Behaviour	Delimitation of responsibilities and transparency Service orientation in working habits Rigor Potential tensions
Organisation	Redundancy free organisation
Skills	New skills correlated to new usages and technology

This example, while simplistic at first sight, shows however that this approach can enable a more consistent and rather fast picture of the enterprise. The next step would consist in defining the baseline state, analysing gaps and defining the transition.

3.2 *Integrating with Current Practices*

While defining the transition, Bob Casey main concern lies in the follow-up phases, especially regarding the integration with current practices in the company, for example enterprise architecture, or skills and change management.

One possible way to deal with his concern is to integrate the approach in existing practices, such as TOGAF. It can indeed be seen as an extension of the analysis in B, C and D phases of the ADM (Architecture Development Method). Starting from phase E, practitioners have the choice between:

a) Restraining themselves to the study of the usual business and IT aspects: in that case this approach would just be an *enhancement of the methodology*
b) Including all other aspects in the transition and planning analysis. In that case, this approach could be seen as a *practice of enterprise transformation* that would embrace other practices than enterprise architecture.

Facing those two options, Bob Casey acknowledged the relevancy of an *enterprise transformation practice*, but due the novelty of the idea, decided to mitigate risks and eventually chose the first option.

3.3 *New Perspectives, New Challenges*

The approach developed and illustrated previously could be a first step to envisage enterprise transformation in a complete, holistic, yet pragmatic manner. It certainly paves the way for new perspectives more consistent with today's environment, as initiated in certain practices such as Praxeme.

Perhaps one day we will see a more efficient and consistent method for transforming the enterprise. Yet, this also leads to new challenges, such as the modelling of all concerns or the integration of all practices within one single framework. There is undoubtedly much work remaining but the effort may be worthwhile to make progress in this highly potential field of enterprise management.

References

1. Gartner: Definition of enterprise architecture, `http://www.gartner.com/it-glossary/enterprise-architecture-ea/` (accessed July 26, 2012)
2. Dijkstra, E.W.: On the role of scientific thought. Selected writings on Computing: A Personal Perspective, pp. 60–66. Springer-Verlag New York, Inc., New York (1982)
3. The Open Group Architecture Framework TOGAF, `http://www.opengroup.org/togaf/` (accessed July 27, 2012)

4. Zachman: Zachman Framework, http://www.zachman.com/ (accessed July 27, 2012)
5. Kotter, J.P.: Leading Change: Why transformation efforts fail. Harvard Business Review, Reprint R0701J (2007)
6. Praxeme: Definition of enterprise architecture, http://www.praxeme.org/index.php?n=Thesaurus.Enterprise Architecture?userlang=en (accessed August 14, 2012)
7. TOGAF: Definition of architecture in TOGAF, http://pubs.opengroup.org/architecture/togaf9-doc/arch/chap03.html (accessed July 27, 2012)
8. Sullivan, A., Sheffrin, S.M.: Economics: principles in action. Pearson Prentice Hall, Upper Saddle River (2003)
9. Skiena, S.S.: The Algorithm Design Manual, 2nd edn. Springer (2010)

Used Abbreviations

IT: Information Technology
TOGAF: The Open Group Architecture Framework
MST: Minimal Spanning Tree
CEO: Chief Executive Officer
SOA: Service Oriented Architecture
ADM: Architecture Development Method

An Enterprise Architecture and Data Quality Framework

Jerome Capirossi and Pascal Rabier

Abstract. Insurance industry undergoes major regulatory changes regarding risk management like Solvency II which require managing data quality. This paper reports an experience feedback about the development of an enterprise architecture and data quality framework suitable for Insurance Industry and COTS[1] environments. The framework, inspired by TOGAF 9.1, is tailored to provide systemic views of enterprise organization, systems and data and to develop joint governance for enterprise architecture and data quality. The paper describes development approach and framework components including metamodel, repository, data quality, tools and governance. it may stand as a proposal for a TOGAF data quality extension.

1 Introduction

Insurance industry sector is undergoing big transformations [1] due to changes in risk regulation approach required by:

- New risks associated with longer life and weather
- Concerns about Asset managers long term solvability
- Financial markets stretching out Insurance Business

Jerome Capirossi
NATEA-Consulting, Charenton le Pont, France
e-mail: jerome@capirossi.org
http://capirossi.org

Pascal Rabier
La Mutuelle Generale, Paris, France
e-mail: prabier@lamutuellegeneral.fr

[1] Commercial off the shelf software.

P.-J. Benghozi, D. Krob, & F. Rowe (Eds.): *Digital Enterp. Des. & Manage. 2013*, AISC 205, pp. 67–79.
DOI: 10.1007/978-3-642-37317-6_7

One of the first releases has been Solvency II directive which led to transformations not as smooth as one may think since Insurance Industry is an old already well structured industry with a fair technical debt. In addition, Insurance risk management relies on data crunching often processed by several stakeholders: brokers, contract managers… That's why data quality is recognized as critical success factor for transparency, good reports and good decision taking.

A mutual insurance company specialized in health and savings has required to extend its capabilities to planning and executing transformations with efficiency. Their today's approach being mainly function by function, does not allow any systemic analyze nor business involvement in projects who only provides requirements and waits for solution. Data quality is also poor since most of systems are COTS bringing their own view on data, and being integrated mainly by data synchronization which appears complex. No existing tools allow to manage data quality especially data controls execution.

2 Approach

The company decided to develop a data quality governance approach. The main idea was the following: in a high Information Technology consuming context, data quality governance without good architecture governance does not make any sense. So it has been decided to develop an architecture and data quality governance joint framework.

Following objectives were set to be enabled in target organization:

- deciding swiftly and efficiently what transformations to do taking in account data quality risks
- addressing efficiently system design questions included data quality risks
- standardizing data quality risk approach in accordance with operational risk approach
- mitigating Information Systems complexity and technical debt
- better controlling and managing subcontractors involvements

A project was launched to customize and roll-out a comprehensive enterprise architecture framework with an extension for data quality which would include strategic and operational levels for governance and development purposes. This framework inspired by TOGAF 9.1[2] would contain content metamodel, repository, some methodological tools and governance organization.

Because it links enterprise architecture and data quality objectives, this framework would be a candidate to apply as a data quality extension of TOGAF 9.1 content metamodel.

3 Metamodel

First step has been to agree on a metamodel allowing 2 modeling levels: macro level on enterprise extent, detailed level for project design. Macro level was intended to shape the boundaries where detailed level has to stay confined. It should be powerful but should remain understandable by stakeholders if we wish they use it. Then, it will be a good basis for governing the whole information system design and data quality.

Metamodel design was conducted in compliance with TOGAF 9.1 and took "Business Service" for pivotal concept.

We define some additional concepts by grouping basic ones. All "Data Entity" related and dedicated to a same topic, "Customer" for instance, were grouped into an "Enterprise Data Entity" concept. All "Business Service" related to a same "Enterprise Data Entity" were grouped into a "Business Service Block". All "Business Service Block" related to a same business function were grouped into a "Business Service Quarter". If some Business Services were not directly related to an "Enterprise Data Entity" but associated to a business function, they were grouped with a "Business Service Quarter" governed by this business function.

Finally, we grouped "Business Service Quarter" into "Business Service Area" according to they are falling into following categories : "Operations" "Operations Support" "Business Intelligence".

We agreed on a 1st model rule which was: only one "Business Service Block" is granted to update a given "Enterprise Data Entity". This meant that if you need to have an up-to-date information from a given "Enterprise Data Entity", you have to use a "Business Service" of its "Business Service Block". Conversely, if you catch an event which requires updating an "Enterprise Data Entity", you need to use a "Business Service" of its "Business Service Block".

We created a dedicated "Business Service Area", called "Master Data", which contains all "Business Service Quarter" associated with "Enterprise Data Entity" which could be qualified as Master Data. They are "Customers", "Products", "Organization", "Contracts", "Persons", "Partners", and "Nomenclatures". Such business services are called from most of systems.

Applications being mainly COTS, most of projects have to deal with integration. For that purpose, we define a pattern which states that each COTS would be viewed as a collection of application components which are grouped by "Business Service Block" called "Logical Application block". Since it supports whole "Business Service Block", it supports all business services and, basically, all data flows provided by the block. Then, main requirement for COTS integration is that interfaces realized by application components should provide data related to the "Enterprise Data Entity" associated with the "Business Service Block".

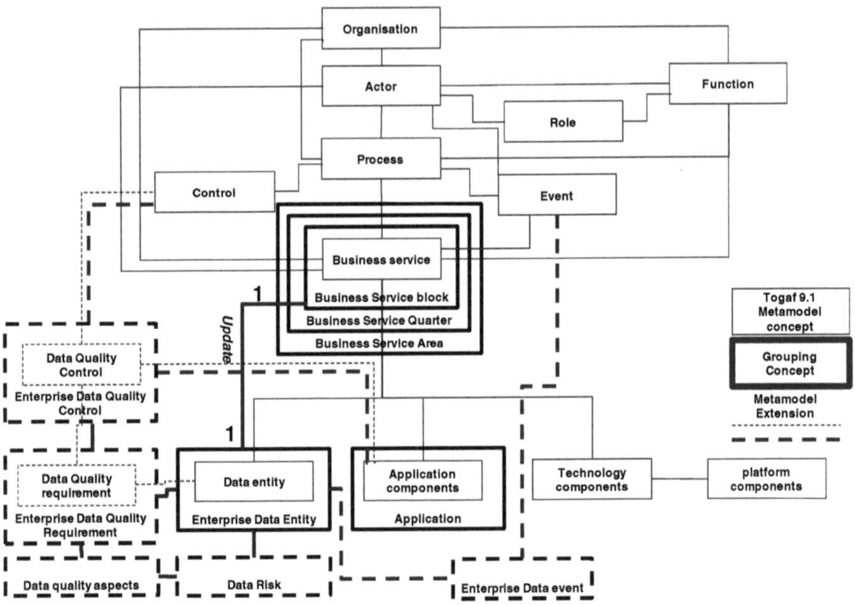

Fig. 1 Metamodel

As each data entity within an "Enterprise Data Entity" would be standardized, COTS would provide a standardized representation of enterprise data which would be aligned with enterprise point of view and not with COTS points of view. In addition, it has the advantage of keeping enterprise away from COTS changes often driven externally.

Same rule would apply to application components since they should get entity data from providers in a standardized form.

Once basic metamodel has been designed, we designed an extension dedicated to data quality management which is detailed in the following paragraph.

4 Data Quality Metamodel Extension

This extension follows same multi-level principle. The lowest basic concept is "Data Quality requirement" which applies to a "Data entity". It is associated to one or several "Data quality control" which allow to check if requirement is met by Data of the "Data Entity".

Data quality requirement should met at best following rules :

- Unambiguously distinct from other requirements,
- It applies to a "Data Entity" or a specific part of "Data Entity"
- It provides value to the business in the mastery of risk of non-quality
- It applies to all instances of a business object or a subset. In this case it will specify the filter criteria (eg contracts whose event "contract")

- It must be tested and verified, it is subject to one or more controls
- It can be weighted with respect to the factor whether it provides fully or partially

Macro level contains "Enterprise Data Quality Requirement" which target all "Data Entity" of an "Enterprise Data Entity". Each "Enterprise Data Quality Requirement" is linked with one or more "Enterprise Data Quality Control" which check that requirement is met at level of "Enterprise Data Entity". In some case, "Enterprise Data Quality Control" may cascade some "Data Quality Control" defined at "Data Quality" level.

In metamodel, data entities are indirectly associated with business events which trigger business services. We added "Enterprise Data Event" concept which group business events directly linked with parts of a given "Enterprise Data Entity" into a kind of macro event. All "Enterprise Data Event" of a given "Enterprise Data Entity" constitutes its lifecycle.

Using the guidelines of International Association for Information and Data Quality[2], we defined data quality aspects [6] suitable for Mutual Insurance business. So, a given "Data Quality Requirement" addresses only one data quality aspect. A given Data quality aspect may be addressed by more than one "Data Quality Requirement". Here after the list of aspects we agreed on:

Table 1 Aspects of Data Quality

Aspect	Description
Accuracy	Data must be properly valued by business transactions.
Consistency	Any data must be consistent with other ones throughout all its life cycle
Uniqueness	No duplicates, either multiple instances of the same data identifier or different identifiers associated with the data.
Integrity	Any data should remain consistent with management rules: regarding mandatory and optional attributes, and other business rules.
Availability	Data must be available whenever user needs to access to it.
Traceability	Data changes are recorded in an audit trail enterprise facility
Completeness	Data must figure all instances of Entities
Compliance	Data representation is consistent with norms and standards of industry or in force in the enterprise
Freshness	Data must reflect the current state is no lag or delay
Intelligibility	Data characteristics and description of different states of life cycle must be obvious and easily understood by stakeholders.

[2] http://iaidq.org

For a given "Enterprise Data Entity", a subset of aspects is selected in accordance with risks attached to data. Indeed, for being mitigated each risk requires some quality aspects to be matched by data. We identify the following risks and their associated aspects:

Table 2 Links between quality risks and aspects

Data risk	Aspect	Data risk	Aspect
Management : bad operations and process supervision	Uniqueness consistency integrity traceability intelligibility	Operational : bad transactional data	Uniqueness consistency compliance availability freshness intelligibility
Contractual : bad bisness contract data	Accuracy uniqueness integrity availability traceability freshness completeness	Intelligence : bad reporting and decision data	Uniqueness consistency integrity availability traceability completeness intelligibility
Financial : bad accounting and finance data	Accuracy consistency integrity availability traceability freshness completeness	Regulatory : bad regulatory reporting data	Accuracy consistency compliance integrity availability traceability freshness completeness

A given "Enterprise Data Quality Control" is executed either by the "Application Component" which produce the dataflow, or other application components or by a business control associated with a process activity as we will see in the following paragraph.

5 Complementary Architecture Patterns

Since business service blocks are tightly linked with enterprise data entities, this metamodel enabled an architecture style which may be called data driven architecture.

In this respect, COTS are viewed as collections of business service blocks which match their functional modules. Pattern states that any COTS dataflow realize logically a remote business service call. An application component called interface executes business service operations and provides data corresponding to the business event. Then any dataflow results from a join of an "Enterprise Data Event" with a "Enterprise Data Entity".

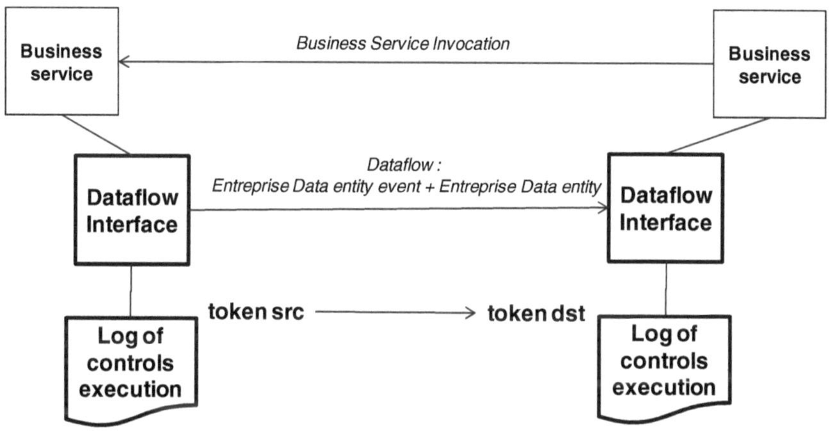

Fig. 2 Data flow reference Architecture

The application component acting as provider is responsible for executing "Enterprise Data Quality Control" and "Data Quality Control", it records execution results in a log associated with an execution token. In this way, all logs could be consolidated to give a crossing view of data quality control results along a business process execution.

Token has been defined to identify any application component interface execution across information system. For enabling end-to-end traceability, token from source is recorded with destination token in consumer application component logging facility.

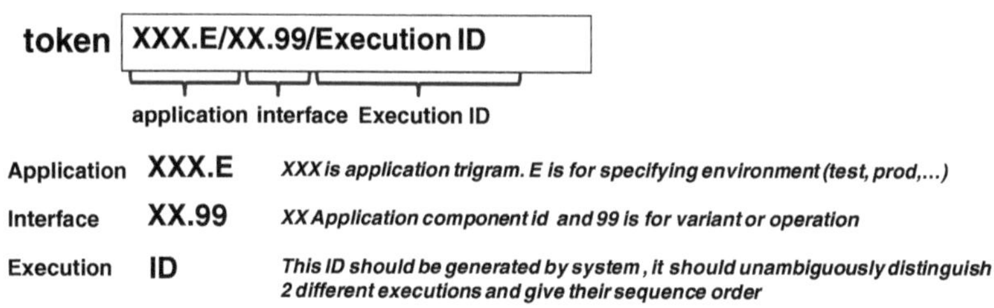

Fig. 3 Token definition

Business service blocks set in "Master Data" business service are realized either by application components of a given COTS which, for some of them, provide directly dataflows to other applications or by a special MDM component which aggregates data from several COTS to create all business services of the block.

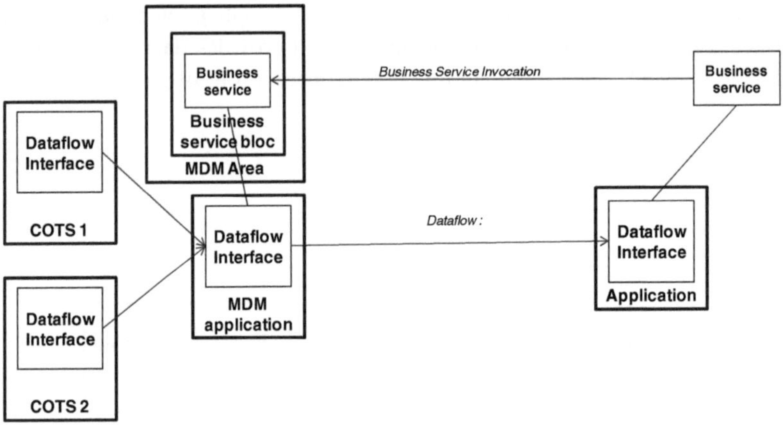

Fig. 4 MDM logical Pattern

In this case control execution and requirements are under the responsibility of the MDM Application components.

6 Tools

Once metamodel and some architecture patterns have been agreed, it has been time to describe architecture landscapes at different levels, macro levels for whole enterprise, detailed level for projects.

Then, enterprise architecture people defined macro level landscapes: process, organization, events, business services, business services blocks, enterprise data entity. These were intended to be used as boundaries and landmarks for projects specifying and developing solutions.

Macro landscapes were done with ARIS from Software AG, but other tools are suitable since they allow to spread architecture landscapes access across all projects stakeholders and architects.

For Data quality, we design a dedicated tool, "Enterprise Data Entity Dictionary" (EDED), which complements landscapes with special information regarding enterprise data entities. We pay attention to be compatible by ISO/IEC 11179 standard [3] related to metadata. Especially we include context information which allows to accept more than one data representation for a couple (event, enterprise data entity) according to the information system context. For example, you may need to have a comprehensive data representation for loading a datawarehouse and a short representation of the same data for a Business transaction.

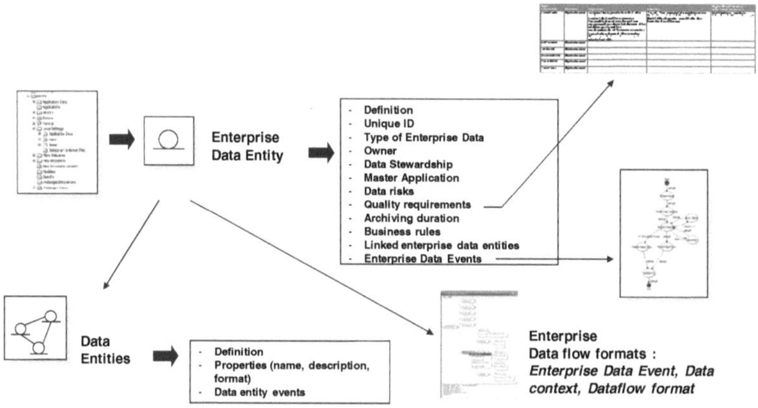

Fig. 5 "Enterprise Data Entity Dictionary" (EDED)

EDED only addresses macro level of architecture, detailed level is covered by regular data dictionaries which usually come with data systems and data modelling tools.

EDED is a repository all information needed by Architects and Data governance. It contains:

- Definition: a shared definition of the Enterprise Data Entity,
- ID: unique identifier by which data is recorded and plotted,
- Type of Enterprise Data Entity: Master data, operation data, intelligence data
- Linked Enterprise Data Entities: the relationships that bind to other Enterprise Data Entities,
- Owner: business owner accountable for quality
- Data stewardship: Business entities that manage operationally data under the guidance of the owner,
- Archiving Duration
- Data Risks
- Enterprise Data quality requirements: with aspects associated to data risks.
- Business Rules: constraints and business rules that must be complied global business object
- Enterprise Data quality events: by the form of finite state diagram

It contains also descriptions of all data entities which constitute "Enterprise Data Entity" under the form of an UML 2.0 class diagram and a table of public data flows formats ordered by event and context. A data flows format is deemed public when it is involved in a dataflow across different enterprise service blocks.

Technical data documentation and dataflows formats are developed by projects and stay under control of IT people. EDED provides only references to retrieve these documents, but does not store them since they are more related to systems than to Enterprise Data.

For Data governance purpose, we design an indicator [5] which could be calculated periodically and support a review by management. It is made of architecture maturity indicators which are viewed as levers for management and operations indicators.

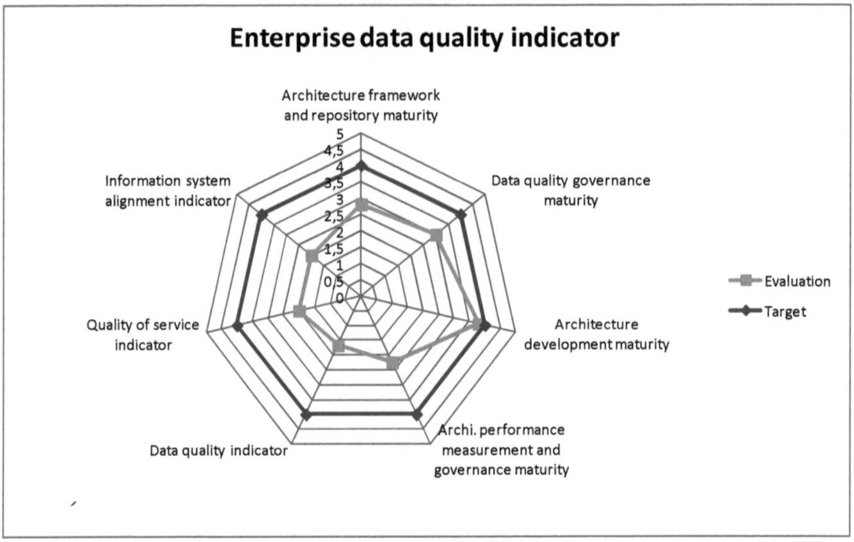

Fig. 6 Enterprise Architecture and Data Quality indicator

Detail is figured below:

- Maturity of Architecture framework and repository: Maintain and spread processes maps, Business services maps, data and application architectures, patterns...
- Maturity of Data quality governance: completion of EDED, effective usage of EDED, performance of Data quality management processes
- Maturity of Architecture development: targets definition and roadmaps, projects scoping, architecture change management
- Maturity of performance measurement and architecture governance
- Consolidated Data quality indicator
- Consolidated quality of service indicator,
- Consolidated information systems alignment indicator to business inspired by COBIT 4.1 [4].

A more detailed version is used by Enterprise architects for managing activity. Then, each indicator is associated with a measurement process, some automated, some others requiring an expert assessment, others a survey.

Table 3 Detailed indicators of architecture and data quality performance

Grouped indicator	Detailed indicators
Maturity of Architecture framework and repository	Completion, compliance and communication of Business architecture maps : organization, processes, services Completion, compliance and communication of Data and Application architecture maps Completion, uptodate and communication of Architecture frame work : processes, pattern, references
Maturity of Data quality governance	Completion, compliance and communication of EDED Effective usage of EDED Performance of Data quality management processes
Maturity of Architecture development	Business and system target and roadmap Enterprise Architecture alignment with strategy Projects compliance with Enterprise Architecture Architecture Requirements Management Architecture change management process Architecture Competences Definition and sourcing
Maturity of performance measurement and architecture governance	Architecture Stakeholders management Architecture Board compliance and performance Project architecture compliance and performance Indicator calculation process compliance and performance
Data quality indicator	Control execution performance Control coverage : executed versus defined Incident due to lack of data quality
Quality of service indicator	% Compliance with Quality of service engagements for each Business domain
Information system alignment	Business alignment detailed indicators: • Compliance with rules regulating • User satisfaction • The IS is a factor of business productivity • The IS is a factor of business security • Frequent and serious incidents in production • Low confidence in IT to achieve projects or improve operations • Ability to integrate business constraints and / or exploit the opportunities of information systems Non functional alignment detailed indicators: • Information system agility • Information system testability • Information system maintainability • Information system easy operations • Information system security • Information system scalability • Information system documentation

7 Architecture and Data Governance

Architecture governance and Data governance rely on the same charter which states that among ultimate goals of Enterprise information systems there is processing data with high quality standards.

Architecture governance main body is the Architecture Board which is accountable for framework change management, architecture development, architecture performance measurement. Mostly, it has been inspired by TOGAF Architecture Board definition.

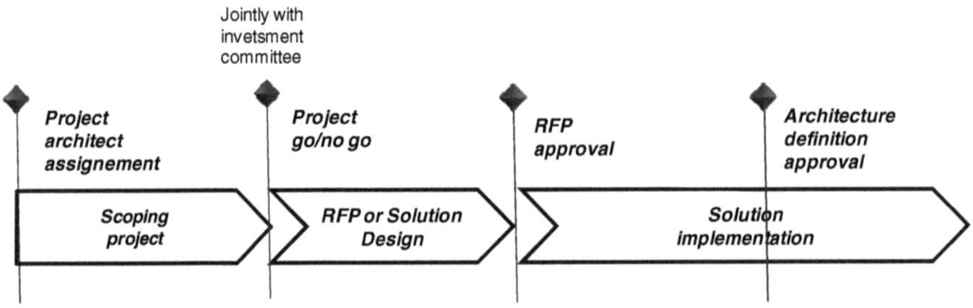

Fig. 7 Architecture Board milestones

It deals with data quality when overseeing architecture development and performance measurement. In architecture development, it approves project architecture definition which include data architecture and data quality requirements.

From a point of view of a given project, all architecture board milestones are undergone submitting the same "Architecture Definition Document" at different stages of development. For helping and supporting projects, we wrote comprehensive guidelines to fulfill the document in compliance with Enterprise architecture standards and objectives and consistent with the company macro architecture. These guidelines make easier integrators management when they are from outside the company.

In performance measurement, architecture board collaborates with data quality governance bodies which provide indicators dimensions directly related to operational data quality.

Data quality governance relies on datastewardship for operational data management and on Data manager for governance. Datastewards perform day to day business information operations, they are using systems and executing controls on data, those specified in EDED[3]. Data manager is a top manager of an enterprise function. He is at stake with data quality for achieving business process performance of business function. He review periodically Architecture and data quality performance indicator. In the case of problems, he requests either action plans to be executed by datatewards or sponsors data quality projects. Then, projects undergo Architecture governance milestone.

[3] Enterprise Data Entity Dictionary.

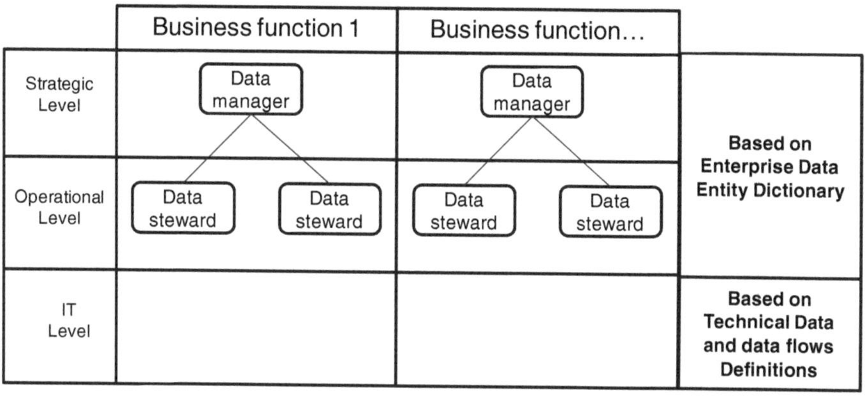

Fig. 8 Architecture Board milestones

The indicator is reviewed periodically by the enterprise board and may be linked with enterprise operational risk management.

8 Conclusion

We rolled-out the framework, coaching projects for developing their Architecture Definition Document. The roll-out was progressive, starting with an experiment on 2 projects, which would be followed by a generalization.

Experimental phase was useful, not only to track implementation bugs, but also to let projects state that framework had simplified architecture definition work. As a result, they became keen to adopt it. Spreading such a message was a good advantage for framework global adoption which is still ongoing.

References

[1] KPMG. Survey - Economic Capital Modeling in the Insurance Industry (August 2012), `http://www.kpmg.com/Ca/en/IssuesAndInsights/ArticlesPublications/Documents/economic-capital-modeling-july-2012.pdf`

[2] Opengroup - TOGAF 9.1 (2010), `http://pubs.opengroup.org/architecture/togaf9-doc/arch/`

[3] ISO/IEC 11179, `http://metadata-standards.org/`

[4] ISACA - COBIT 4.1 (2007), `http://www.isaca.org/`

[5] Thorn, S.: When was your las enterprise architecture maturity assessment – Open group Blog (March 2012), `http://blog.opengroup.org/2012/03/02/when-was-your-last-enterprise-architecture-maturity-assessment/`

[6] Wang, R.Y., Strong, D.M.: Beyond accuracy: What data quality means to data consumers. Journal of Management Information Systems 12, 4 (1996)

Aligning Alignment with Strategic Context: A Literature Review

Kari Hiekkanen, Mika Helenius, Janne J. Korhonen, and Elisabete Patricio

Abstract. The alignment of business and IT has been a persistent topic of discussion in the past decades. As information systems have evolved from an administrative support function to an integral part of business fabric, the classic "internal" perspective adopted by the bulk of alignment research falls short in accounting for the dynamic business network context and continuous evolution with the environment. The information systems planning and strategy discourse should transcend the notion of "alignment" and bring out the strategy-shaping role of IT. This paper presents a classification of business–IT alignment approaches vis-à-vis respective schools of thought in strategic management. Both disciplines are seen to co-evolve with the increasingly complex "strategic context". The approach is meant to help contextualize extant and future work in terms of underlying assumptions and thereby make more conscious statements about the practical applicability of research topics, methods and results in varying contexts. As relatively simple, static and mechanistic conceptualizations of strategy and business–IT alignment render inadequate, concepts such as dynamic capabilities, co-evolution and organizational ambidexterity represent a more adaptive and more encompassing approach to make sense of the increasingly complex strategic context.

1 Introduction

One of the enduring themes in information systems planning and strategy is the *alignment* of business and IT. The business–IT alignment, or strategic alignment, is commonly viewed as a desired and important factor and driver of optimizing

Kari Hiekkanen · Mika Helenius · Janne J. Korhonen · Elisabete Patricio
Aalto University, School of Science
Department of Computer Science and Engineering
Otakaari 1, 02150 Espoo, Finland
e-mail: `{firstname.lastname}@aalto.fi`

P.-J. Benghozi, D. Krob, & F. Rowe (Eds.): *Digital Enterp. Des. & Manage. 2013*, AISC 205, pp. 81–98.
DOI: 10.1007/978-3-642-37317-6_8

business performance. The impact of alignment on business performance has been studied for several decades (e.g. [37][51][9][91][45][79][53][54][55]).

The notion of alignment has its roots in the "design school" of strategy, where the essential components are the extent of congruence, or fit, between an organization's internal structure and its external situation [61]. Strategy is concerned with a match between internal resource capability and external opportunity towards superior performance [73]. In contemporary business environment, where organizations need to be innovative, flexible and faster due to uncertainty, complexity and change of the "environment", the complex and diverse nature of strategy renders the concept of alignment increasingly problematic. Strategy is no longer a "big idea" for many companies as business environment is far different – calling for new means to conduct and contextualize strategy [35]. Strategy is seen more as an emergent [63] and continuous practice based process [100].

In digital enterprises, where the marketplace is global and interconnected, discontinuities such as technological breakthroughs, new regulations, and geopolitical upheavals are frequent and non-linear. The competitive advantage is in constant flux and organizations are forced to find ways to reinvent their very essence without falling apart [34]. The advances in technology both enable and drive firms to change their business models. Digitalization and networked information economy have brought unprecedented changes to markets and business models, disrupting entire industries [7]. Phenomena, such as disintermediation and reintermediation [29], digital goods [74], dematerialization and liquification of resources [66], and new types of technology-mediated interactions brought by the Internet [15] characterize the digital enterprise. Information systems have evolved from administrative, functionally oriented support systems to an integral part of business fabric that is fused into products and services.

The classic "internal" perspective adopted by the bulk of alignment research falls short in accounting for the dynamic business network context and continuous evolution with the environment. We view that the information systems planning and strategy discourse should transcend the notion of "alignment" and its associated connotations of "business–IT divide" and "IT follows business". It should rather acknowledge the strategy-shaping role of IT (cf. [18]). Thus, alignment as the underlying concept of IT management and governance frameworks and practices needs to be reviewed in the context of contemporary perspectives on strategy and strategizing. In line with the recent observation that the traditional notion of strategic fit has possibly lost its explanatory power [104], we concur that organizations should defer from focusing too much on either efficiency or flexibility and rather develop dynamic capabilities [90] that enable a more balanced, ambidextrous [68] behavior between exploitation and exploration [59].

Motivated by Leonard's [50] call for exploring alternative approaches to alignment, this article attempts to outline the evolution of alignment discourse vis-à-vis relevant strategic management concepts in order to provide a better understanding of the assumptions and perspectives on strategizing that underlie previous alignment research. Our aim was not to conduct an exhaustive analysis but merely suggest linkage points between strategic management approaches on one hand, and various approaches to business–IT alignment on the other hand, in an attempt to uncover ontological assumptions underlying alignment research toward strategy and strategizing. The approach is meant to help contextualize extant and future work in terms of underlying assumptions and thereby make more conscious statements about the practical applicability of research topics, methods and results in varying contexts, We view that this reconceptualization would help identify and manage IT-based competencies and capabilities in digital enterprises where IT is a core business asset.

The article is organized as follows. First, we outline the evolution of business–IT alignment concepts. In conducting the review, we followed the systemic approach suggested by Webster and Watson [97]. The aims of a systematic review can be varied and include: (1) clarifying the relative strengths and weaknesses of the literature on the question, (2) summarizing a fairly large amount of literature, (3) resolving literature conflicts, (4) avoiding a redundant unnecessary case, and (5) improving the generalizability of literature findings. Our aim is to provide a contemporary view of the previous work and highlight various gaps by analyzing relevant literature. We first analyze "major contributions" in the field; secondly review "backward" and "forward" cited articles. We acknowledge that our review is limited, but it should provide a relevant coverage of the field. Also, we have limited our review to publications in English language only. After providing an overview of the extant alignment research, we compare the models with corresponding strategic management literature and based on our interpretative understanding of the ontological assumptions in selected models. Thereby, we identify the general tone of the alignment discourse throughout time, the underlying assumptions, as well as respective approaches to strategic management. Finally, we uncover underexplored fields in alignment research and chart out possible future directions for alignment discussion based on a more synergistic, ambidextrous concept of alignment.

2 Alignment Research – An Overview

The term align originates from the French word *ligne* meaning "line" and the Latin word *linea* meaning "string". It has the following meanings: a) to bring into line or alignment b) to array on the side of or against a party or cause (transitive verb) or a) to get or fall into line b) to be in or come into precise adjustment or correct relative position (intransitive verb) [92]. Thus, the notion of alignment suggests a sequential execution from strategy to IT.

In their extensive bibliographical study, Chan and Reich [20] summarized 150 different articles on alignment, spanning three decades of research in the field. The articles use several terms for alignment such as fit, linkage, integration, coherence, harmony, fusion, congruence and variation. These are all used for alignment, although some minor differences in their use exist. The term business–IT alignment also takes different forms in the literature, and can be written as business/IT alignment, business and IT alignment, business–IT alignment, IT alignment, and alignment of business and IT, all meaning the same. Also the terms IT, ICT and IS are often used interchangeably.

Several models of alignment have been proposed by adopting the organization view [48]. The early approaches to alignment include alignment coordination model [49], fit [96] and forces interaction [56]. MacDonald [56] and Baets [2] were among the first ones to associate a process view to alignment. As the technological development lead to a wider adaptation and use of IT, the tension of new technology choices [39] induced seeking for balance between alignment [37], linkage [77], and harmony [103].

In line with Henderson's and Venkatraman's [37] strategic alignment model (SAM), the bulk of alignment research builds on the principle of separation between business and IT domains with a number of variable elements, such as organizations, plans, processes, competences etc. The SAM model is probably the most widely adapted model of alignment and it has been studied from the empirical perspective (e.g. [14][1]) and also extended by other researchers (e.g. [52][57][58][1]). More recent studies have approached strategic alignment from the perspectives of resource-based view [42][43] and dynamic capabilities [76][21][28][4] attempting to bridge the "gap" between IT, alignment and strategy research.

Several dimensions of alignment are discussed in literature including strategic, intellectual, structural, social and cultural [22]. The strategic dimension focuses on the complementary aspects of business and IT strategies and plans, including aspects of strategic information systems planning. The structural alignment dimension focuses on the structural fit between business and IT decision-making structures and organizations. The role of informal structures (relationships and communication) in alignment success has also been discussed in literature [19]. The social dimension is defined as the state in which business and IT executives within an organizational unit understand and are committed to the business and IT mission, objectives, and plans [78].

The result of three decades of alignment studies has brought us an astonishing set of partly competing, partly overlapping approaches, models and frameworks. There is a steady growth in the number of academic papers on alignment and the main bulk of research consists of work developing new instrumental support artifacts for alignment [44]. On one hand, the pluralism is the strength of the field: different perspectives and disciplinary contributions provide far more insights into the relationship between business and IT than any single perspective could do. On the other hand, the proliferation of models, concepts and frameworks fosters complexity in which it seems easy to get lost. From the practitioner's point of view, there is a challenge in knowing *which* model to apply, *when* and *how*.

Apparently, academic research on alignment has provided little practical value to organizations. Previous arguments to this phenomena point to models, which are not feasible to apply, which were developed conceptually, and that do not derive from the real world [18]; validated results are not concise, and models are prone to subjectivity [104][1]. Other arguments for the lack of value refer to overly mechanistic models, which are unsuitable for contemporary organizations [38].

The mechanistic approaches do not account for organizations as organic, dynamic, and ambiguous wholes, with relationships that are parallel and simultaneous [93]. Many approaches also omit the formal and informal roles of participants – e.g. people – in organizations. Leavitt's [48] argument that organizations could be usefully viewed as complex socio-technical systems, comprising four elements (objectives, structure, technology, and people), is overlooked in many models.

Already in the 90's, Ciborra [17][18] points out that much of the alignment discussion naïvely assumes that enterprise reality can be captured objectively and can be controlled and made predictably via linear cause and effect chains. He further questions the implicit dominance of a structured strategy process in an era when uncertainty and flexibility predominate and when the articulation of the strategic intent is difficult. This poses a significant challenge, because most alignment models presuppose an existing business strategy to which an IT organization can align itself [21].

There is a general agreement that organizations with "high" alignment outperform those with "lower" alignment of business and IT. Tallon and Kraemer [86] found a positive and significant relationship between strategic alignment and IT payoffs, but they also uncovered evidence of an alignment paradox: beyond a certain critical point, further increases in strategic alignment lead to lower IT payoffs. Especially so-called dual focus firms are "forced to rethink any move that involves an increase in strategic alignment if at the same time this could lead to a reduction in the payoffs they realize from their IT investment". Short-term IT support for the business strategy may limit organizational flexibility and prevent the organization from responding to the changes in the environment at some future point.

In a similar vein, Sphilberg et al. [81] maintain that an organization that aligns IT well with business, but is not effective, tends to fall in an "alignment trap", where IT spending is in increase but growth is slow in coming. Sphilberg and his colleagues found that for the majority of high-performing organizations that are both highly aligned and highly effective, the path has been that of first increasing the effectiveness of the IT organization, while temporarily forgetting about enhancing alignment. This may require changing the alignment perspective from that of traditional strategy execution to an appropriate alternative (cf. [37]).

3 Alignment and Strategic Management

The evolution of modern approaches to strategic management can be characterized by the dichotomy between two research streams: *strategy content* and *strategy process*. Content research seeks to answer the question of *what* constitutes competitive advantage; process research is concerned with *how* strategies emerge over time [60]. The former research stream seeks to understand the relationship between strategy and performance using a structural approach to industries and competitive forces, whereas the latter stream is about descriptive studies of how strategies are formed and implemented.

In aligning alignment approaches with respective schools of thought in strategic management as presented in Table 1, we have used this division between content (*what*) and process (*how*) as a guiding principle in lining up respective approaches and streams. Few alignment approaches explicitly base their arguments on a certain strategic management theory. As some of the selected alignment models

Table 1 Strategic Management Viewpoints and Alignment Approaches

	Strategic Management	**Respective Alignment Approaches**
Content-Based Stream		
Market-Based View	Hedley 1977 [36] Porter 1980 [73]	Henderson and Venkatraman 1993 [37] Maes 1999 [57] Bergeron, Raymond and Rivard 2004 [10]
Resource-Based View, Knowledge-Based View	Wernerfelt 1984 [99] Barney 1991 [5] Peteraf 1993 [72] Grant 1996 [32] Sveiby 2001 [84]	Kearns and Lederer 2003 [46] Peppard and Ward 2004 [71] Kearns and Sabherwal 2006 [47]
Dynamic Capabilities	Teece et al. 1997 [89] Eisenhardt and Martin 2000 [25] Benner and Tushman 2003 [8] Teece 2007 [90]	Sun and Chen 2006 [83] Chen et al. 2008 [23] Gogan et al. 2010 [31] Baker et al. 2011 [4]
Ambidexterity	March 1991 [59] O'Reilly and Tushman 2007 [69] Gibson and Birkinshaw 2004 [30]	Sabherwal et al. 2001 [80]
Process-Based Stream		
Strategy as Process	Mintzberg 1973, 1978 [62][63] Johnson 1987 [43] Burgelman 1986, 1991 [11][12] Moncrieff 1999 [65]	Baets 1992, 1996 [2][3] Burn 1996 [13] Reich and Benbasat 2000 [78] Peppard and Breu 2003 [70] Benbya and McKelvey 2006 [6]
Strategy as Practice	Whittington 1999, 2003, 2006 [100][101][102] Jarzabkowski 2003, 2005 [40][41] Vaara and Whittington 2012 [94]	Ciborra 1997 [18] Galliers 2006, 2007, 2011 [26][27][28] De Vaujany 2008 [95]

include both *structural* and *process* elements, our assessment is based on an interpretative understanding of the focus or "*the center of gravity*" of each approach and the tone of the discussion by the authors.

3.1 Content-Based Stream

Porter [73] points out that competition goes beyond established industry rivals to include four other competitive forces as well: customers, suppliers, potential entrants, and substitute products. However, this market-based view of strategy is not interested in the resources businesses have and treats their behavior as a "black box". Competitive strategy determines how the organization gains an advantage over its rivals within chosen market positions [73]. Although these strategic choices are numerous, the environment is assumed as relatively stable and major changes (e.g. disruptive technologies, market upheavals) as infrequent.

Classic *structural* approaches (e.g. [37][57]) to business–IT alignment presume an external strategy to *align to;* the relationship to business strategy is more sequential, following the "IT follows business" mindset, and the focus is more on *what* needs to be aligned but there is far less consensus on *how* the alignment is to be achieved [50].

In parallel to the market-based view, other studies switched their focus from industry structure as a unit of analysis to that of the organization's internal structure, resources and capabilities. According to the resource-based view (RBV) [99][5][72], asymmetries in the resources and capabilities of businesses in the same industry are the source of competitive advantage. To sustain this competitive advantage, the resources need to be valuable, rare, inimitable and non-substitutable [5]. The knowledge-based view (KBV) is similar to RBV, but instead of a broad range of resources as the basis of corporate strategy, the knowledge-based view focuses on a particular type of resource – knowledge. Knowledge is seen "as the most strategically important of the firm's resources" [32].

From the RBV perspective, strategic IT alignment can create competitive advantage, when it represents a complex organizational process that is both heterogeneous and immobile [46]. The process of strategic IT alignment is a capability in itself and advantage occurs when IT is used to leverage the organization's resources in some inimitable way (ibid.). When alignment is seen through the lens of the resource-based view, value comes not from replication but from uniqueness [85]. Knowledge-based view on alignment [47] concentrates on the knowledge-based theory [32] linking knowledge considerations to strategic alignment and business effects of IT.

In more dynamic markets, however, resource fortification of the RBV can be problematic. The focus of competition is shifting from the management of internal resources to selecting and developing technologies and business models that build competitive advantage, through assembling and orchestrating difficult-to-replicate co-specialized assets [89]. The dynamic capability approach (e.g. [89][25][8][90]) focuses on how organizational and strategic management competencies can enable organization to explore, exploit and capture market opportunities in order to achieve and sustain competitive advantage in an open, rapidly changing environment [8][90]. The dynamic capability perspectives on alignment focus on adapting, integrating, and reconfiguring skills, resources and abilities, and view alignment process as a dynamic capability that reconfigures specific IT assets to support other core resources ([83][23][31][4]).

March [59] observed that organizations tend to concentrate either on capabilities for exploitation or exploration. Exploitation focuses on activities and behaviors that improve the performance of the current business, whereas exploration aims at ensuring the future effectiveness of the business. Exploitation is about efficiency, increasing productivity, control, certainty and variance reduction; exploration is about search, discovery, autonomy, innovation and embracing variation [68]. However, a dominant focus on either exploitation or exploration may result an undesired situation for the organization [75][104].

An organization that is able to simultaneously explore and exploit is called *ambidextrous* [68]. Recent studies on organizational ambidexterity [30][75][76][104] show that organizations that achieve a high-level balance between both exploitation and exploration are more successful than organizations that focus only on either set of capabilities. From a strategic management perspective, the punctuated equilibrium viewpoint of alignment by Sabherwal et al. [80] can be considered as ambidextrous as it corresponds to the cyclical domain in the typology of Simsek et al. [82], in which ambidexterity is achieved through sequential allocation of resources on relative stable exploitation interspersed by sporadic episodes of quick exploration and change.

3.2 Process-Based Stream

The alignment models corresponding to the process stream of strategic management research focus on the dynamism of business–IT alignment, the co-evolutionary development of both strategy and IT strategies and on the social dimension of alignment. These models highlight the importance of the process in which internal politics, organizational culture, managerial cognition and skills help achieve and maintain high alignment. The central theme is that alignment is perceived mainly as a dynamic, ongoing process and not as a conceivable end-state.

More recently, the strategy-as-practice approach to strategic management [100] depicts strategy as an activity undertaken by people, not as a formal property of organizations. From an epistemological point of view, the strategy-as-practice approach understands practice as being "closer" to reality and delivering a "more accurate" description of the real world phenomena than formal theories populated by multivariate analyses of firm or industry-level factors. The strategy-as-practice approach is very much couched in European characters and is clearly to be understood as a systematic critique of orthodox, hegemonic, and mainly North American-inspired strategy research [16].

In line with this strategy-as-practice viewpoint, Ciborra [17][18] argues that serendipity and improvisation (e.g. "tinkering" and "bricolage") are more likely to yield competitive advantage from information systems than deliberate planning of the type that is generally prescribed when seeking strategic alignment. He emphasizes the role of praxis and notes that organizations that consistently pursue IS innovation are more likely to have unique capabilities developed over time, through experience or tinkering with multifarious technologies, that enable them to quickly assess the potential of emerging technologies to contribute to their business strategy.

De Vaujany [95] argues that multilayered, multifaceted nature of IS strategic value is shaped and reshaped by the intra- or extra-organizational praxis of some leading actors originating the value. In this view, the focus of management should be on IS strategic potential, IS realized values, and final economic performance, rather than on business–IT alignment. All this should then be considered as a complex "system in practice".

Galliers [26][27][28] focuses also more on the process of strategizing rather than on the outcome of the process. He argues that benefit is to be gained from a more inclusive, exploratory approach to the strategy process. He further proposes a strategizing framework facilitating modes of exploitation and exploration. The process of exploitation adopted in the framework bears many of the hallmarks of mainstream thinking on the IS strategy, and in the exploration aspect the emphasis is much more on issues associated with situated learning, communities of practice and cross-project learning. The idea is to accommodate both deliberate and emergent modes of strategizing.

4 Discussion

In this paper, we conducted a literature review and put forth a classification of business–IT alignment approaches by respective schools of thought in strategic management. Our aim was not to conduct an exhaustive analysis but merely suggest linkage points between strategic management approaches on one hand, and various approaches to business–IT alignment on the other hand, in an attempt to uncover ontological assumptions underlying alignment research toward strategy and strategizing. The approach is meant to help contextualize extant and future work in terms of underlying assumptions and thereby make more conscious statements about the practical applicability of research topics, methods and results in varying contexts. More specifically, if the complexity of the strategic context, de facto, precludes certain paradigmatic approaches to strategic management (e.g. the market-based view), it also rules out respective stances of business–IT alignment.

It is to be noted that many other ways to categorize both strategic management and alignment approaches could be used as an analytical lens. One alternative approach would be for example utilize the typology of ten strategic management schools by Mintzberg et al. [64] as a base of analysis.

In their research concerning ambidexterity and fit in strategic management, Wulf et al. [104] concluded that ambidexterity is a much better predictor of organizational performance than the concept of strategic fit. They argue that top management should defer from a too focused alignment of the organization on either efficiency or flexibility. Instead the management should aim to develop capabilities for ambidexterity to ensure sustained high performance.

In line with Galliers [27][28], we subscribe to the holistic approach in which both exploitation and exploration are accounted for. Also, recent research by Tallon and Pinsonneault [87] notes the usefulness of ambidexterity in thinking about alignment and agility in IS research.

Examining presented alignment models through the taxonomy presented by Simsek et al. [82], the punctuated equilibrium model by Sabherwal et al. [79] is comparable with *cyclical ambidexterity*. For other dimensions of ambidexterity – *reciprocal, partitional* and *harmonic* – the compatibility of extant alignment models is debatable.

In the *reciprocal* dimension, ambidexterity is achieved through the efficient specialization of exploitation and exploration across intra- or inter-organizational network where different strategic stances can be pursued sequentially across different participants.

In the *partitional* dimension, ambidexterity is achieved through structural partitioning of the pursuit of exploitation and exploration into separate units each having its own strategies and structures [8].

From the alignment perspective, both *reciprocal* and *partitional* ambidexterities present a number of challenges, such as *what* to align, to *whose* strategies to align to and *who* maintains the balance between different perspectives. Alignment problems are especially relevant in these dimensions if organization in question pursues centralized or federal IT governance archetypes [98].

From the alignment perspective, the *harmonic* ambidexterity, the concurrent pursuit of exploitation and exploration harmoniously within the same organizational unit is probably hardest to accommodate to. This dimension is inherently challenging as simultaneous pursuit can lead to conflicts, contradictions and inconsistencies in all areas, including IT. This happens since pursuing harmonic ambidexterity becomes intertwined in both strategic and operational activities of the unit's culture, structure and systems [12]. In this dimension, alignment approaches based on strategy as practice perspective are probably more suitable as this dimension places a premium on individual's learning and integrative abilities in line with Ciborra [18] and Galliers [27][28].

The digital enterprises with virtual value chains can be described as complex adaptive business systems, where the competitive performance landscapes of products and services are highly dynamic and co-evolve. This challenges the notion of "alignment" as the question of overarching strategy has become difficult to answer. Consequently, aligning IS strategy with competitive strategy alone might offer limited and inconsequential results [88].

Sustainable competitive advantage of a complex adaptive business system requires that organizations co-evolve within the dancing, rugged competitive landscape. Agility in adaptation to the changing environment is vital but not enough. Given the long lead times and costs entailed in the development and deployment of IT capabilities, the core assumptions and models that become embedded in the IT capabilities tend to structure the actions of organizations and to remain relatively static over a long period of time. Structures embedded in IT pose risks for the organization's attempts to co-evolve [88].

Galliers [27] also rises a question of how to align a relatively fixed IT that is implemented in an organization with a business strategy and associated information requirements that are in constant need of adjustment. He names the lack of dynamism as one of the core problems with alignment and calls for flexible – or agile – IT. Galliers also argues that some organizations, in pursuing efficiency and reducing costs through IT, may have lost agility in the process.

Oh and Pinsonneault [67] note that trying to sustain "perfect" alignment may be an illusionary concept, given the speed and magnitude of changes affecting organizations. They posit that organizational complexities hinder organizations from perceiving the true consequences of misalignment. Just as a small environmental change can cause a significant impact on the sustainability of alignment, a small change in alignment can result in a dramatic consequence for organizational performance. They conclude that continuous refinement and fine-tuning are necessary to maintain superior organizational performance.

Doz and Kosonen [24] discuss how to enable business model renewal and how to make an organization more agile by developing three core meta-capabilities: strategic sensitivity, leadership unity and resource fluidity. Strategic sensitivity pertains to "the sharpness of perception of, and the intensity of awareness and attention to, strategic developments." As such, it likens to Teece's *sensing* of new opportunities [89][90]. Leadership unity, in turn, can be seen corresponding to *seizing* opportunities (ibid.); it is about integrated and fast decision-making by the top management to consolidate pertinent prospects. Resource fluidity refers to the internal capability to reconfigure capabilities and rapidly redeploy them. Again, apparent analogy to Teece's *reconfiguration* (ibid.) can be seen.

We maintain that relatively simple, static and mechanistic conceptualizations of strategy and business–IT alignment are being challenged by more adaptive and more encompassing views. The dynamic capabilities approach, co-evolutionary views and the concept of ambidexterity can be seen as representative of approaches to make sense of the increasingly complex, technology-induced strategic context.

5 Conclusions

After reviewing the literature on business–IT alignment and recent developments in the field of strategic management, we view that the current alignment discussion is still biased towards a mindset, in which IT is seen as a separate, value-adding function, whose focus is on present-day value realization, operational quality and reliability rather than as the source of strategic advantage.

Most business–IT alignment models adopt a static, mechanistic and segmented worldview on organizations and technology, in which alignment is conceived as known, quantifiable, achievable and measureable [38]. In digital enterprise settings, however, thinking alignment in mechanistic terms may be limiting at best, and fatal at worst, as current "status-quo" of business and IT is constantly being challenged by new realities, non-linear discontinuities and incessant technological development. The classic alignment models based on the notion of competitive strategy [73] or resource-based view [99][5][72] are not sufficient in the agile, networked and complex business environment that calls for new cognitive and systemic capabilities in leadership, technology and processes. A narrow focus of alignment underestimates the systemic complexity of IT that addresses different business needs [81].

As the role and importance of information and information systems in contemporary organizations increase, new perspectives are needed in managing, operating and innovating IT-based business models. The agile, networked and complex business environment of today calls for the extension of existing management principles and practices to embrace higher levels of complexity and multifaceted nature of alignment and adaptive capabilities. New concepts and theories that can provide the genesis of a new management paradigm are needed [33].

Accordingly, we view that the future of business–IT alignment research and discussion should be based on more contemporary notions of strategic advantage. One such development would be to embrace the notion of dynamic capabilities and the ambidextrous forms of organizations. Since IT is an extension of strategy in contemporary organizations, the alignment discussion and models should reflect the notion of strategic ambidexterity – the balance between exploitation and exploration capabilities. However, since exploiting existing competences and exploring new opportunities involve contradictory capabilities, the question is how organization can achieve and maintain the balance between these orientations. Possible solutions proposed by previous research in strategic management highlight the role of structural, contextual and leadership-based solutions in achieving ambidexterity.

The next step would be to extend the tentative results of this literature review with empirical analysis on how organizations have achieved organizational ambidexterity and how this is reflected in information systems planning and strategy practice; whether extant alignment approaches are suited to ambidextrous organizations; and how to asses business-IT alignment in a such contexts?

The developmental requirements of the executives open up a research avenue in its own right: what "dynamic leadership capabilities" are required from business and IS leaders to create and run ambidextrous organizational forms and carry out business model change?

References

[1] Avison, D., Jones, J., Powell, P., et al.: Using and validating the strategic alignment model. Journal of Strategic Information Systems 13, 223–246 (2004)

[2] Baets, W.: Aligning Information Systems with Business Strategy. Journal of Strategic Information Systems 1(4), 205–213 (1992)

[3] Baets, W.: Some Empirical Evidence on IS Strategy Alignment in banking. Information & Management 30(4), 155–177 (1996)

[4] Baker, J., Jones, D., Cao, Q., et al.: Conceptualizing the Dynamic Strategic Alignment Competency. Journal of Association for Information Systems 12(4), 299–322 (2011)

[5] Barney, J.: Firm Resources and Sustained Competitive Advantage. Journal of Management 17(1), 99–120 (1991)

[6] Benbya, H., McKelvey, B.: Using coevolutionary and complexity theories to improve IS alignment: a multi-level approach. Journal of Information Technology 21, 284–298 (2006)

[7] Benkler, Y.: The Wealth of Networks: How Social Production Transforms Markets and Freedom. Yale University Press, New Haven (2006)

[8] Benner, M., Tushman, M.: Exploitation Exploration And Process Management: The Productivity Dilemma Revisited. Academy of Management Review 28(2), 238–256 (2003)

[9] Bergeron, F., Raymond, L.: The contribution of information technology to the bottom line: a contingency perspective of strategic dimensions. In: Proceedings of the International Conference on Information Systems, pp. 167–181 (1995)

[10] Bergeron, F., Raymond, L., Rivard, S.: Ideal patterns of strategic alignment and business performance. Information & Management 41, 1003–1020 (2004)

[11] Burgelman, R.: A Model of Interaction of Strategic Behaviour Corporate Context and the Concept of Strategy. The Academy of Management Review 8, 61–70 (1989)

[12] Burgelman, R.: Intraorganizational ecology of strategy making and organizational adaptation: Theory and field research. Organization Science 2, 239–262 (1991)

[13] Burn, J.: IS Innovation and Organizational Alignment – A professional juggling act. Journal of Information Technology 11(1), 3–12 (1996)

[14] Burn, J., Szeto, C.: A comparison of the view of business and IT management on success factors for strategic alignment. Information and Management 37, 197–216 (2000)

[15] Burnett, R., Marshall, P.: Web Theory: An Introduction. Routledge, New York (2003)

[16] Carter, C., Clegg, S., Kornberger, M.: Strategy as Practice? Strategic Organizations 6(1), 83C–99C (2008)

[17] Ciborra, C.: From thinking to tinkering. In: Ciborra, J. (ed.) Strategic Information Systems: a European Perspective. Wiley, Chichester (1994)

[18] Ciborra, C.: De Profundis? Deconstructing the Concept of Strategic Alignment Scandinavian. Journal of Information Systems 9(1), 57–82 (1997)

[19] Chan, Y.: Information Systems Strategy Structure and Alignment. In: Papp, R. (ed.) Strategic Information Technology: Opportunities for Competitive Advantage, 1st edn., pp. 56–81. Idea Group Publishing, Hershey (2001)

[20] Chan, Y., Reich, B.: IT alignment: an annotated bibliography. Journal of Information Technology 22, 316–396 (2007)
[21] Chan, Y., Reich, B.: IT alignment: what have we learned? Journal of Information Technology 22, 297–315 (2007)
[22] Chan, Y., Reich, B.: Rethinking Business-IT Alignment. In: Galliers, R., Currie, W. (eds.) Management Information Systems, Critical Perspectives and New Directions. Oxford University Press, New York (2011)
[23] Chen, R., Sun, C., Helms, M., et al.: Aligning information technology and business strategy with a dynamic capabilities perspective: A longitudinal study of a Taiwanese Semiconductor Company. International Journal of Information Management 28, 366–378 (2008)
[24] Doz, Y., Kosonen, M.: Embedding Strategic Agility: A Leadership Agenda for Accelerating Business Model Renewal. Long Range Planning 43(2-3), 370–382 (2010)
[25] Eisenhardt, K., Martin, J.: Dynamic capabilities: What are they? Strategic Management Journal 21(10-11), 1105–1121 (2000)
[26] Galliers, R.: On confronting some of the common myths of Information Systems strategy discourse: towards a revised framework. In: Mansell, R., et al. (eds.) Oxford Handbook on Information and Communication Technologies. Oxford University Press, Oxford (2006)
[27] Galliers, R.: Strategizing for Agility: Confronting Information Systems Inflexibility in Dynamic Environments. In: Desouza, K. (ed.) Agile Information Systems: Conceptualization Construction and Management. Butterworth-Heinemann, Burlington (2007)
[28] Galliers, R.: Further Developments in Information Systems Strategizing: Unpacking the Concept. In: Wendy, C., Galliers, R. (eds.) The Oxford Handbook of Management Information Systems: Critical Perspectives and New Directions. Oxford University Press, Oxford (2011)
[29] Giaglis, G., Klein, S., O'Keefe, R.: Disintermediation, Reintermediation, or Cybermediation? The Future of Intermediaries in Electronic Marketplaces, Global Networked Organizations. In: Proceedings 12th Electronic Commerce Conference, Modernaorganizacija (1999)
[30] Gibson, C., Birkinshaw, J.: The antecedents consequences and mediating role of organizational ambidexterity. Academy of Management Journal 47(2), 209–226 (2004)
[31] Gogan, J., Lewis, M., Sankaranaryanan, B., et al.: Aiming at a Moving Target: IT Alignment in Toy Companies. In: 18th European Conference on Information Systems (2010)
[32] Grant, R.: Toward a knowledge-based theory of the firm. Strategic Management Journal 17, 109–122 (1996)
[33] Grant, R.: The Future of Management: Where is Gary Hamel Leading Us? Long Range Planning 41, 469–482 (2008)
[34] Hamel, G., Välikangas, L.: The quest for resilience. Harvard Business Review 81(9), 52–63 (2003)
[35] Hamel, G.: Strategy Innovation and the Quest for Value. Sloan Management Review 39, 7–14 (1998)
[36] Hedley, B.: Strategy and the Business Portfolio. Long Range Planning 10(1), 9–15 (1977)

[37] Henderson, J., Venkatraman, N.: Strategic Alignment: Leveraging information technology for transforming organizations. IBM Systems Journal 32(1), 4–16 (1993)

[38] Hoogervorst, J.: Enterprise Governance and Enterprise Engineering. Springer, London (2009)

[39] Ives, B., Järvenpää, S., Mason, R.: Global business drivers: aligning information technology to global business strategy. IBM Systems Journal 32(1), 143–161 (1993)

[40] Jarzabkowski, P.: Strategic Practices: An Activity Theory Perspective on Continuity and Change. Journal of Management Studies 40(1), 23–55 (2003)

[41] Jarzabkowski, P.: Strategy as practice. Sage, London (2005)

[42] Johanssen, B.: Get There Early: Sensing the Future to Compete. Present Berrett-Koehler Publishers (2008)

[43] Johnson, H.: Strategic Change and the Management Process. Basil Blackwell, Oxford (1987)

[44] Kaidalova, J., Seigerroth, U.: An inventory of the business and IT alignment research field. In: Abramowicz, W., Domingue, J., Węcel, K. (eds.) BIS Workshops 2012. LNBIP, vol. 127, pp. 116–126. Springer, Heidelberg (2012)

[45] Kearns, G., Lederer, A.: The Effect of Strategic Alignment on the Use of IS-based Resources for Competitive Advantage. Journal of Strategic Information Systems, 265–293 (2000)

[46] Kearns, G., Lederer, A.: A Resource-Based View of Strategic IT Alignment: How Knowledge Sharing Creates Competitive Advantage. Decision Sciences 34(1), 1–29 (2003)

[47] Kearns, G., Sabherwal, R.: Strategic Alignment Between Business and Information Technology: A Knowledge-based View of Behaviours Outcomes and Consequences. Journal of Management Information Systems 23(3), 129–162 (2006)

[48] Leavitt, H.: Applied organizational change in industry: Structural technological and humanistic approaches. In: March, J. (ed.) Handbook of Organizations, Rand McNally, Chicago, pp. 1144–1170 (1965)

[49] Lederer, A., Mendelow, A.: Coordination of Information Systems Plans with Business Plans. Journal of Management Information Systems 6(2), 5–19 (1989)

[50] Leonard, J.: What are we Aligning? Implications of a Dynamic Approach to Alignment. In: ACIC 2008 Proceedings (2008)

[51] Luftman, J., Lewis, P., Oldach, S.: Transforming the Enterprise: The alignment of business and information technology strategies. IBM Systems Journal 32(1), 198–221 (1993)

[52] Luftman, J., Papp, R., Brier, T.: Enablers and Inhibitors of Business-IT Alignment. Communications of the Association for Information Systems 1 (1999)

[53] Luftman, J.: Assessing Business Alignment Maturity. Communications of AIS 4(14) (2000)

[54] Luftman, J., Kempaiah, R.: An Update on Business-IT Alignment: 'A Line' has been drawn. MIS Quarterly Executive 6(3), 165–177 (2007)

[55] Luftman, J., Ben-Zvi, T., Dwivedi, R., et al.: IT Governance: An Alignment Maturity Perspective. International Journal of IT/Business Alignment and Governance 1(2) (2010)

[56] MacDonald, H.: Business strategy development alignment and redesign. In: Scott-Morton, M. (ed.) The Corporation of the 1990s – Information Technology and Organizational Transformation. Oxford University Press, New York (1991)

[57] Maes, R.: A Generic Framework for Information Management. Primevera Working Paper (1999)

[58] Maes, R., Rijsenbrij, D., Truijens, O., et al.: Redefining business-IT alignment through a unified framework. Landelijk Architectuur Congres, Amsterdam (2000)

[59] March, J.: Exploration and Exploitation. Organizational Learning Organization Science 2(1), 71–87 (1991)

[60] Mellahi, K., Sminia, H.: Guest Editors' Introduction: The frontiers of strategic management research. International Journal of Management Reviews 11(1), 1–7 (2009)

[61] Miles, R., Snow, C.: Organizational Strategy, Structure and Process. McGraw-Hill, New York (1978)

[62] Mintzberg, H.: The Nature of Managerial Work. Harper and Row, New York (1973)

[63] Mintzberg, H.: Patterns in Strategy Formation. Management Science 24(9), 934–948 (1978)

[64] Mintzberg, H., Ahlstrand, B., Lampel, J.: Strategy Safari: Your complete guide through the wilds of strategic management, 2nd edn. Prentice Hall, UK (2009)

[65] Moncrieff, J.: Is strategy making a difference? Long Range Planning Review 32(2), 273–276 (1999)

[66] Normann, R.: Reframing Business: When the Map Changes the Landscape. Wiley, Chichester (2001)

[67] Oh, W., Pinsonneault, A.: On the assessment of the strategic value of information technologies: conceptual and analytical approaches. MIS Quarterly 31(2), 239–265 (2007)

[68] O'Reilly, C., Tushman, M.: The ambidextrous organization. Harvard Business Review 82, 74–81 (2004)

[69] O'Reilly, C., Tushman, M.: Ambidexterity as a Dynamic Capability: Resolving the Innovator's Dilemma. Working Paper # 07-088, Harvard Business School, Cambridge MA (2007)

[70] Peppard, J., Breu, K.: Beyond Alignment: A coevolutionary view of the information systems strategy process. In: Twenty-Fourth International Conference on Information Systems (2003)

[71] Peppard, J., Ward, J.: Beyond strategic information systems: towards an IS capability. Journal of Strategic Information Systems 13, 167–194 (2004)

[72] Peteraf, M.: The Cornerstones of Competitive Advantage: A Resource-Based View. Strategic Management Journal 14(3), 179–191 (1993)

[73] Porter, M.: Competitive Strategy. Free Press, New York (1980)

[74] Quah, D.: Digital Goods and the New Economy, CEPR Discussion Paper No. 3846 (2003)

[75] Raisch, S., Birkinshaw, J.: Organizational ambidexterity: Antecedents outcomes and moderators. Journal of Management 34(3), 375–409 (2008)

[76] Raisch, S., Birkinshaw, J., Probst, G., et al.: Organizational ambidexterity: Balancing exploitation and exploration for sustained performance. Organization Science 20(4), 685–695 (2009)

[77] Reich, B.: Investigating the Linkage between Business and Information Technology Objectives: A multiple case study in the insurance industry. Doctor of Philosophy Thesis, University of British Columbia (1992)
[78] Reich, B., Benbasat, I.: Factors that influence the social dimension of alignment between business and information technology objectives. MIS Quarterly 24(1), 81–113 (2000)
[79] Sabherwal, R., Chan, Y.: Alignment Between Business and IS Strategies: A Study of Prospectors Analyzers and Defenders. Information Systems Research 12(1), 11–33 (2001)
[80] Sabherwal, R., Hirschheim, R., Goles, T.: The Dynamics of Alignment: Insights from a Punctuated Equilibrium Model. Organization Science 12(2), 179–197 (2001)
[81] Shpilberg, D., Berez, S., Puryearm, R., et al.: Avoiding the Alignment Trap in IT. MIT Sloan Management Review 49(1) (2007)
[82] Simsek, Z., Heavey, C., Veiga, J., et al.: A Typology for Aligning Organizational Ambidexterity's Conceptualizations Antecendents and Outcomes. Journal of Management Studies 46(5) (2009)
[83] Sun, C., Chen, R.: A study on the strategic alignment process with information technology for new ventures: from a dynamic capability perspective. In: ECIS 2006 Proceedings Paper 202 (2006)
[84] Sveiby, K.: A knowledge-based theory of the firm to guide in strategy formulation. Journal of Intellectual Capital 2(4), 344–358 (2001)
[85] Tallon, P.: A Process-Oriented Perspective on the Alignment of Information Technology and Business Strategy. Journal of Management Information Systems 24(3), 227–268 (2007)
[86] Tallon, P., Kraemer, K.: Investigating the Relationship between Strategic Alignment and IT Business Value: The Discovery of a Paradox. In: Shin, N. (ed.) Creating Business Value with Information Technology: Challenges and Solutions. Idea Group Publishing, Hershey (2003)
[87] Tallon, P., Pinsonneault, A.: Competing Perspectives on the Link Between Strategic Information Technology Alignment and Organizational Agility: Insights from a Mediation Model. MIS Quarterly 25(2), 463–486 (2011)
[88] Tanriverdi, H., Rai, A., Venkatraman, N.: Reframing the Dominant Quest of Information Systems Strategy Research for Complex Adaptive Business Systems. Information Systems Research 21(4), 822–834 (2010)
[89] Teece, D., Pisano, G., Shuen, A.: Dynamic Capabilities and Strategic Management. Strategic Management Journal 18(7), 509–533 (1997)
[90] Teece, D.: Explicating dynamic capabilities: the nature and microfoundations of (sustainable) enterprise performance. Strategic Management Journal 28(13), 1319–1350 (2007)
[91] Teo, T., King, W.: Assessing the impact of integrating business planning and IS planning. Information and Management 30, 309–321 (1996)
[92] The Oxford English Dictionary, 2nd edn. Oxford University Press, Oxford (1989), http://www.oed.com/ (accessed May 10, 2012)
[93] Tsoukas, H.: New Thinking in Organizational Behavior. Butterworth-Heinemann, Oxford (1994)
[94] Vaara, E., Whittington, R.: Strategy as Practice: Taking social practices seriously. Academy of Management Annals (2012)

[95] De Vaujany, F.: Strategic Alignment: What Else? A Practice Based View of IS Value. In: Proceedings of Twenty Ninth International Conference on Information Systems, Paris (2008)

[96] Venkatraman, N.: The concept of fit in strategy research: toward verbal and statistical correspondence. Academy of Management Review 14(3), 423–444 (1989)

[97] Webster, J., Watson, R.: Analyzing the past to prepare for the future: Writing a literature review. MIS Quarterly 26(2), xiii–xxiii (2002)

[98] Weill, P., Ross, J.: IT Governance: How top performers manage IT decision rights. Harvard Business School Press, Boston (2004)

[99] Wernerfelt, B.: The Resource-Based View of the Firm. Strategic Management Journal 5(2), 171–180 (1984)

[100] Whittington, R.: Strategy as Practice. Long Range Planning 29(3), 731–735 (1999)

[101] Whittington, R.: The work of strategizing and organizing: for a practice perspective. Strategic Organization 1(1), 117–125 (2003)

[102] Whittington, R.: Completing the Practice Turn in Strategy Research. Organization Studies 27(5), 613–634 (2006)

[103] Woolfe, R.: The Path to Strategic Alignment. Information Strategy 9(2), 13–23 (1993)

[104] Wulf, T., Stubner, S., Blarr, H.: Ambidexterity and the Concept of Fit in Strategic Màngement – Which Better Predicts Success? HHL Working Papers, Leipzig Graduate School of Management (2010)

Digital Value Chains for Carbon Emission Credits

Ichiro Satoh

1 Introduction

The reduction of greenhouse gases (GHGs), including carbon dioxide (CO_2), has become one of the most serious issues in the world today to build a sustainable world. Digital economy have several contributions to the reduction of GHG emissions. Among them, *Carbon credits* is one of the most important and effect approach to reducing the amount of GHG emissions, where carbon credits are generated by the reduction of CO_2 emissions in sponsoring projects, which increase CO_2 absorption, such as renewable-energy, energy-efficiency, and reforestation projects. Although carbon credits themselves do not reduce the amount of CO_2 emissions around the world, they are important incentives for GHG reduction projects.

Many companies have also sold products with the amount of carbon credits equivalent to the amount of GHGs emitted due to the use or disposal of products so that the credits have been used to offset GHGs emissions. There are a variety of products on the market with carbon credits, e.g., automobiles, disposable diapers, and toys. For example, from September 2007, Lufthansa began offering its customers the opportunity of offsetting carbon emissions through voluntarily donating carbon credits to mitigate the amount of CO_2 emitted due to the actual average fuel consumption per passenger.

However, carbon offsetting poses several problems that result from carbon credit trading. Carbon credits are usually acquired through *carbon credit trading* between countries or companies, or in markets via professional traders, called carbon traders or agencies. Furthermore, existing trading schemes are performed through e-commerce, but they are too complicated for non-professional traders, individuals, or small and medium-sized enterprises, to participate in. In fact, the minimal unit of

Ichiro Satoh
National Institute of Informatics
2-1-2 Hitotsubashi, Chiyoda-ku, Tokyo 101-8430, Japan
e-mail: ichiro@nii.ac.jp

P.-J. Benghozi, D. Krob, & F. Rowe (Eds.): *Digital Enterp. Des. & Manage. 2013*, AISC 205, pp. 99–110.
DOI: 10.1007/978-3-642-37317-6_9

existing credit trading is usually more than one hundred or one thousand tonnes of CO_2 emissions, whereas the amount of GHGs emitted due to the use or disposal of consumables is less than one kilogram.

To solve problems in existing carbon trading, this paper aims at proposing a new scheme for trading carbon credits as a new digital economic system for reducing carbon emissions on the earth by using infrastructure and RFID tags (or barcodes). It enables a small amount of carbon credits attached to products to be transferred to endconsumers who buy these products and carbon credits to be easily traded. The key idea behind our proposed approach is to use RFID tags (and barcodes) as certificates for the rights to claim carbon credits, because RFID tags are widely used in the management of supply chains. Another idea is to use the return of RFID tags as an authentication mechanism. We designed a digital architecture for managing RFID-enabled carbon credit offsetting and trading. The architecture was constructed and evaluated with real carbon credits in a real supply-chain system.

2 Background

Before describing our approach, this section briefly outlines carbon credits and their associated problems. Carbon credits are tradable certificates, also called *carbon emission credits* or *carbon offset credits*, that represent a certain volume of absorbed or reduced emissions by different people or organizations that have reduced excessive amount of GHGs in the atmosphere in the short- or long-term. For example, developed countries or companies financially or technically support projects that aim to reduce GHG emissions in developing countries. They can, in turn, offset their emissions by credits generated from the projects. These projects might involve installing renewable energy technologies, implementing energy efficiency measures, or removing CO_2 from the atmosphere through carbon sequestration.

There are other tradable emission rights, called *carbon emission caps* (or *carbon (emission) allowances*), that are limits. A government authority first sets limits on the amount of CO_2 that companies are allowed to emit. If a company emits an amount of CO_2 due to its activities below its limit, it can sell the excess capacity, which is the difference between the limit and the amount of CO_2 that really has emitted, as caps to companies whose emissions are over their limits. If a company emits an amount of CO_2 beyond its limit, it must pay a penalty or buy caps from someone to comply with its caps. Carbon emission caps are traded between companies or in markets.

Our approach supports both carbon emission credits and intends to treat carbon emission credits in a unified manner so that it can assign carbon emission credits into products and claim the credits in a unified approach. Therefore, in the following sections, we simply describe both carbon emission credits and carbon emission caps as carbon credits to avoid any misunderstanding between them.

3 Problem Statements

This paper describes a digital economic system for trading carbon emission credits, which have several properties that other economic values do not have.

Difficulty in Carbon Credit Trading:

Carbon credits can usually be traded through electronic commerce systems, but existing systems are quite different from digital enterprise. They are closed and complicated so that only professional traders, called *carbon providers* or *carbon agencies*, buy or sell carbon credits on behalf of their clients. Therefore, it almost impossible for end-users, small companies, or NPOs/NGOs to sell or buy carbon credits.

Large Trading Units:

The minimal units of existing trading carbon emission credits, including caps, are more than one hundred or one thousand tonnes of CO_2 . However, the amount of CO_2 an average person emits throughout his/her life for one year is less than one tonne. Each endconsumer product should have less than one kilogram of carbon credits to offset CO_2 emissions in the use or disposal of the product. However, there are currently no approaches to trading small amount of carbon credits, e.g., one gram or one kilogram.

Non-voluntary Carbon Offsetting Products:

As existing schemes for carbon offsetting products any mechanisms to transfer the carbon credits linked to the products to endconsumers. Instead, dealers or manufacturers, who assign the credits to products, offset the credits on behalf of the purchasers (buyers) of the products. Although the purchaser pay extra corresponding to the credits, they have no chance of owning the credits and they do not know whether the credits have been properly offset by the dealers or manufacturers.

4 Carbon Emission Trading Approach

This paper proposes a new digital enterprise approach for enabling a small amount of carbon credits to be attached to each item and to be transferred to consumers who buy it. Our approach introduces RFID tags (or barcodes) as carbon credits for the rights to claim credits in carbon offsetting, because RFID tags (or barcodes) are used in supply chains. In fact, our approach can use the RFID tags that have already been attached to items for supply chain management. The approach was designed as a complement to existing supply management systems. It therefore has nothing to do with the commerce of items themselves. It also leave the transfer of carbon credit between companies with existing carbon trading systems, because commerce for carbon credits must be processed by certificated organizations. Instead, the approach is responsible for attaching carbon credits to RFID tags and claims for carbon credits. The approach should support emission credits in a unified manner. It also should not distinguish between items for end-consumer and others, because non-end-consumers may buy items for end-consumers. Some readers may think that our approach is trivial. However, simplicity and clarity are essential to prompt most people and organizations to participate or commit to activities to reduce GHG emission by carbon offsetting.

Our approach satisfies the following main requirements: 1) The approach needs to encourage industries and homes to reduce GHG emissions. It also needs to be compliant with regulations on carbon offsetting. 2) Simplicity must be a key concern in minimizing operation costs, because it tends to be in inverse proportion to cost. This is needed for people and organizations to understand what is required of them. 3) Any commerce scheme provides the potential to advantage some participants at the expense of others. The approach enables organizations or people that reduce more GHG emissions to be rewarded with greater advantages. 4) The values of carbon credits, particular emission credits tend be varied. The amounts, expiration dates, and sources of all carbon credits, which may be attached to items, need to be accessible. 5) When consumers purchase items with carbon credits, they should easily be able to own the credits without any complicated operations to authenticate them. 6) Item commerce in the real world is often done in warehouses and stores, where networks and electronic devices may not be available. Our approach itself should be available offline as much as possible.

5 Digital Architecture for Carbon Offsetting and Trading

Our approach introduces RFID tags (or barcodes) as carbon credits for the rights of emitters to claim credits in carbon offsetting and trading, because RFID tags (or barcodes) are used in supply chains. In fact, our approach can use the RFID tags that have already been attached to items to manage supply chains. The approach was designed to complement existing supply management systems. It therefore has nothing to do with the trading of the items themselves. It also leaves the transfer of carbon credits between companies to existing carbon trading systems, because carbon trading must be processed by certificated organizations. Instead, the approach is responsible for attaching carbon credits to RFID tags and claims for carbon credits. The approach should support emission credits in a single manner. It should also not distinguish between items for endconsumers and others, because non-endconsumers may buy items for endconsumers.

5.1 RFID Tags as Certificates to Claim Carbon Credits

The basic idea behind our approach is to use RFID tags themselves, rather than their identifiers, as certificates for carbon credits. If the people or organizations that claim the credits with only the identifiers, we need a complicated authenticate them. It is difficult to replicate or counterfeit RFID tags whose identifiers are the same, because their identifiers are unique and embedded into them on the level of semiconductors.

To claim carbon credits dominated by RFID tags, we need to return these RFID tags to the stakeholders that assigned carbon credits to the tags. This is because there is at most one RFID tag whose identifier is the same. RFID tags can be used as certificates for carbon credits. For example, when sellers want to attach carbon offsetting credits to items, they place RFID tags on them that represent the credits for the items. Our approach couples carbon credits with RFID tags themselves, instead of the identifiers of the RFID tags. Therefore, purchasers, who buy the items, tear the RFID tags from them and return the tags to the sellers (or the stakeholders of the credits). When the sellers receive the RFID tags from the purchasers, they transfer the credits to any accounts for payments that the purchasers specify.

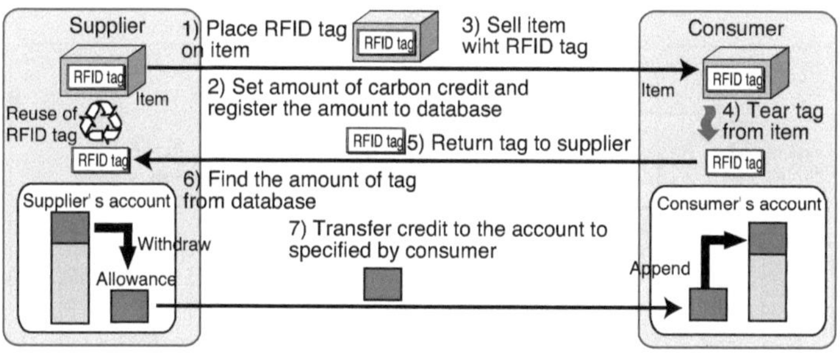

Fig. 1 RFID-based attachment of carbon credits to items

Figure 1 explains our approach to attach carbon credits to items with RFID tags, which involves seven steps

1) A seller places an RFID tag on an item (or a volume of items) if the item has no tag.
2) It sets a certain amount of carbon credits for offsets for a item and registers the amount and the identifier of the tag in a database.
3) It sells the item with the RFID tag to a purchaser.
4) The purchaser tears the tag from the item that it has bought.
5) It only returns the tag with information about the account that the credit should be paid to, to the seller.
6) The seller receives the tag and then finds the amount of carbon credits coupled to the tag in the database.
7) It transfers the amount to the account specified by the purchaser and removes information on the identifier from the database so that the tag can be reused.

5.2 *Carbon Credit Trading with RFID Tags*

When a purchaser has torn an RFID tag from an item, which might have been attached to an item that he/she purchased, our approach permits the purchaser to resell the tag to others (Figure 2). Instead, the new holder of the tag can claim the carbon credits attached to the tag from the stakeholder of these credits or resell them to someone else. Note that trading RFID tags corresponds to trading carbon credits.

To offset GHG emissions according to the Kyoto protocol, we must donate certified carbon credits to the government via a complicated electronic commerce system. Our approach provides two approaches to carbon offsetting. The first is to simply donate RFID tags coupled to certified carbon credits to the government. For example, people can simply throw RFID (unsigned) tags into mailboxes to contribute to reducing GHG emissions in their home countries. The government then gathers the posted tags. The second is to explicitly specify the certificated cancellation account of the government as the account that the credit should be paid into.

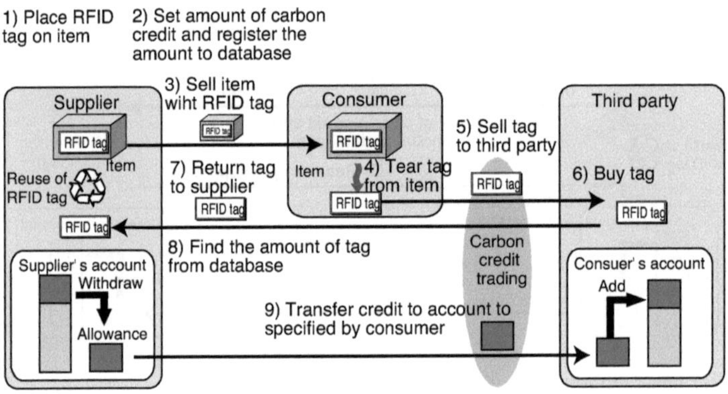

Fig. 2 RFID-enabled trading of carbon credit

6 Implementation

Our approach has been designed so that it can be used in supply chains, where each supply chain consist of sellers, retailers, distributors, transporters, storage facilities, and customers who are involved in moving an item or service from upstream to downstream. Our approach assumes that sellers at steps in a supply chain will sell their items to customers, including raw materials and components, with RFID tags coupled to carbon credits.

- Our approach requires each RFID tag to have its own unique read-only identifier. Most RFID tags used in supply chain management already have such identifiers. For example, we can directly use RFID tags that have been standardized by EPCglobal, because their identifiers, called Electronic Item Code (EPC), consist of the identifiers of the dealers, sellers, manufacturers, or other agents, and the identifiers of the individual items themselves.
- Anyone can access information about the credits attached to the items, because the credits are transferred to purchasers who return the tags themselves to the sellers. The sellers should provide information about the credits, e.g., their amounts, expiration dates, and sources.

- To support carbon offsets, the amount of credits attached to a item need to be equivalent to the total or partial amount of CO_2 emissions resulting from the use or disposal of the items. Nevertheless, the approach itself is intended to leave the amount of credits attached to an item at the stakeholder's discretion, because the credits can be an incentive to sell the item.

Some readers may worry that returning RFID tags to their stakeholders is more costly than returning the identifiers of tags via a network. There are two flows that are opposite to each other between sellers and purchasers at each stage in real supply chains; the flows of items and the flows of receipts or containers for the items. Our approach can directly use the latter flow to return tags from purchasers to sellers. Therefore, our cost and extra CO_2 emissions are small. Actually, returnable containers or boxes, which deliver parts or components from sellers and then return them to sellers, are widely used in real supply chains. Our approach can introduce carbon credits as incentives to return such returnable materials to their stakeholders.

To store carbon credits in a certified manner, many companies entrust certified agents to store and transfer their carbon credit accounts just like they entrust their money to banks. Our system assumes that sellers (and purchasers) have such agents. However, existing agents for carbon credit accounts are not concerned with carbon credits that are RFID tag-based. To solve this problem we introduced a new organization, called carbon credit RFID tag agents (simply called RFID agents), which is not included in existing schemes for carbon offsetting and trading. It is responsible for managing RFID tags and carbon credits coupled with the tags. Figure 3 is a minimal set in our system between a seller and a buyer.[1] Each subsystem has four kinds of facilities.

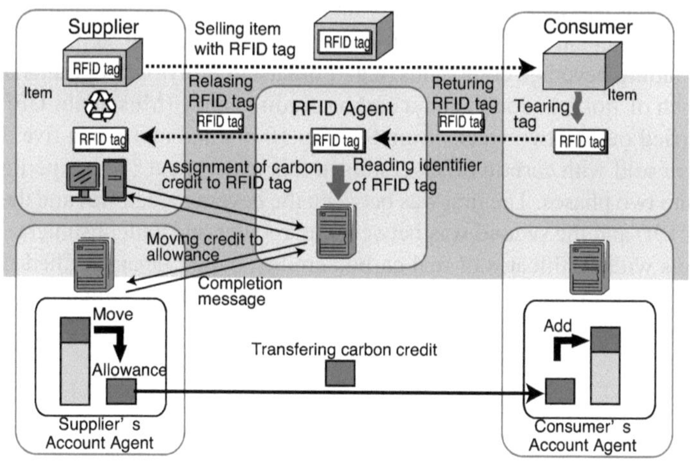

Fig. 3 System architecture

[1] The proposed approach presented in this paper is described by supports gray rectangular parts in the figure.

- Each seller has at least one carbon credit account entrusted to agents for carbon credit accounts. It has RFID tag reader systems to read the identifiers of RFID tags. If a seller consigns one or more RFID agents to manage RFID tags for carbon credits, they need a database to maintain which RFID agent will manage each of the RFID tags.
- Each purchaser may have at least one carbon credit account entrusted to agents for carbon credit accounts. It buys items that RFID tags have attached to them for carbon credits from sellers or traders. It needs RFID tag reader systems, when it intends to access information about carbon credits.
- Agents for carbon credit accounts, simply called *account agents*, may be existing certified carbon providers. They have two databases. The first maintains carbon credit accounts and the second maintains information about assigned credits. They can only be connected to certain RFID agents and other account agents through authenticated and encrypted communications.
- An RFID agent has a database to couple the identifiers of RFID tags and information about carbon credits. The agent may lease RFID tags, which may already have been assigned a certain amount of credits to sellers.

In the following explanation, we have assumed that RFID tags have been provided to sellers by RFID agents and the identifiers of RFID tags contain the identifiers of agents in addition to the unique identifiers of items. The system in Figure 3 is self-contained but it may cascade from upstream to downstream along a supply chain.

7 Social Experiment

The experiment was an early case study on the proposed approach, but was carried out on a supply chain for beverages it was evaluated at several steps in the supply chain, including beverage companies (e.g., Pokka and Fujiya), a supermarket (Kitasuna branch of Ito-yokadou)[2], and a carbon credit agency (Mitsubishi UFJ Lease). It was carried out for two weeks from 9 am to 10 pm and more than five thousand goods were sold with carbon offset credits in this experiment.[3] We experiment was divided into two phases. The first was between the beverage factories and the retailer (supermarket) and the second was between the retailer and endconsumers. We provided items with certificates of real carbon credits, called Japan Verified Emission Reduction (J-VER), where J-VER credits were generated from thinning forest and were traded on the domestic market and managed by the Forestry Agency.[4]

[2] The supermarket is one of the biggest in Tokyo area.

[3] We have left the results of user evaluation to a future paper, because here we have aimed at presenting our basic ideas and implementation rather than user evaluations.

[4] We initially acquired 6 tonnes of J-VER carbon emission credits in the experimental and divided them into small amounts, e.g., 300, 500, or 700 g, and assigned them to goods.

7.1 *Carbon Credits Attached to Factory Items*

The beverage factories filled cans with beverages and packed twenty or thirty cans into returnable containers or cardboard boxes.[5] They adhered RFID-tags onto these returnable containers and placed attachable RFID-tags onto the cardboard boxes. They shipped such containers or boxes to retailers. Each RFID tag represented a specified amount of J-VER carbon credits. The supermarket bought the containers or cardboard boxes with the RFID tags attached. It tore the tags from the cardboard boxes and returned them to the factories so that it could claim the credits to be transferred to it. It returned the returnable containers to the factories without tearing the RFID tags. The return rate of RFID tags attached to containers was about 80 %. We knew that carbon credits coupled to the tags were an incentive for the containers to be returned to the factories (Figure 4). There were no extra operations to return RFID tags that had been adhered to the returnable containers, because the tags were returned with the containers to the factory. Retailers disposed of the cardboard boxes instead of returning them. Nevertheless, since they returned the acceptance notification documents of items or additional order documents to the factories, the RFID tags attached to the cardboard boxes could be sent with such documents to the factories.

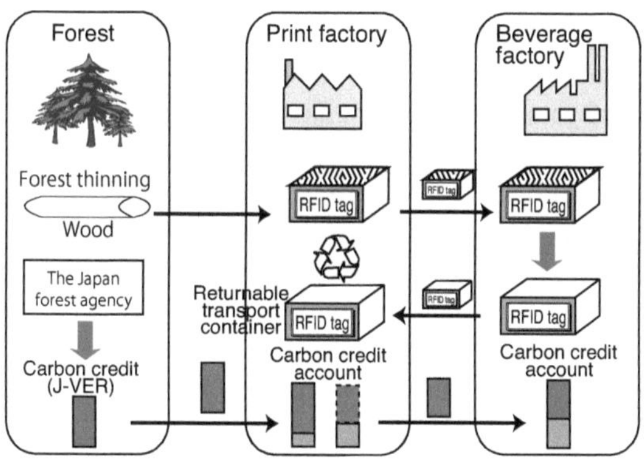

Fig. 4 Experiment in supply chain

7.2 *Carbon Credits Attached to Final Products*

The supermarket opened the returnable containers or cardboard boxes containing the cans. It attached a barcode seal on the cans and sold them to endconsumers, where each barcode seal displayed small amount of J-VER carbon credits, because the price of each RFID tag was relatively more expensive than the price of a can.[6] Each

[5] The cans, called *CartoCans*, were made almost entirely of paper processed from wood chips obtained from thinning forests.

[6] A can costs less than one dollar and an RFID tag seal costs more than a dime.

barcode was formatted in a 2D barcode, called QR code, and consisted of its own identifier, the weights of carbon emission credits assigned to it, and the address of the management server. Figure 5 shows beverage cans with barcodes in a showcase at the supermarket.

Endconsumers bought cans and collected barcode seals as their carbon credits. We supported two cases to reclaim credits in the proposed approach.

- The first was for endconsumers to return barcode seals to the supermarket to reclaim credits. Cashiers could distinguish between original or imitation seals, because they received the seals themselves. Therefore, even when someone read the barcodes attached to the cans, the endconsumers who bought the cans could reclaim the credits.
- The second was for endconsumers to read barcodes by using their own scanners, e.g., cellular phones with cameras, and they then sent the information to the server specified in the barcode. As some might peel off the barcode attached on the cans in stores and illegally reclaim carbon credits. The barcodes were concealed by covering them with other seals.[7]

We used the former in our approach. Few endconsumers participated in the latter, because most endconsumers wanted to immediately reclaim their credits. The former also enabled consumers to access information about carbon credits by reading the barcodes with cellular phones before they bought the items attached with the barcodes. The experiment assumed that retailers bought barcode seals that had already been assigned to small amounts of carbon emission credits, like postage stamps. This is because small retailers might not have any terminals.

In the experiment, we provided the supermarket with barcode seals that had 300, 500, and 700 g weights of carbon credits. The supermarket sold 5320 cans, which consisted of twenty-one kinds that barcode seals had attached according to their environment load. The sales volumes of cans with carbon credits in two weeks was three times more than usual at the supermarket. Thirty-five percent of barcodes were returned to the supermarket by customers who claimed the credits. The experiment enabled consumers to offset their CO_2 emissions by using the carbon credits they reclaimed from the barcodes.

7.3 Lessons Learned

There were many lessons learned from the experiment. Most problems in the experiment were not technical. For example, many consumers asked us about the notion of carbon offsets and credits so that we spent a lot of time to dealing with their questions. The current implementation was not built for performance, but we measured the cost of attaching the carbon credits to an RFID tag after the tag's identifier had been read, which was 460 ms. The cost of claiming carbon credits was 390 ms. The costs tended to increase according to the number of simultaneous requests.

[7] The seals that concealed the barcodes used special adhesive. They could not be attached to the barcode after they had been torn.

Fig. 5 Beverage with Barcode for carbon credits

Nevertheless, our system was designed to handle attachment and claim requests through batch processing. The number of attachment and claim requests was also bound to the number of RFID tags or barcodes.

8 Related Work

Several researchers have explored computing technology to make a contribution to the environment. For example, Persuasive Appliances [6] was an interface system to provide feedback on energy consumption to users. PowerAgent [1] was a game running on mobile phones to influence everyday activities and minimize the use of electricity in the domestic settings. UbiGreen [3] was an interactive system running on mobile phones and gave users feedback about sensed and self-reported transportation behaviors to reduce CO_2 emissions from the transportation sector.

There have been several projects that have used sensing systems to manage warehouses and logistics to reduce CO_2 emissions. Ilic et al. [4] proposed a system for controlling the temperature of perishable goods to reduce GHG emissions. Dada, et al. [2] proposed a system for accurately quantifying GHG emissions to calculate carbon footprints and communicate the results to consumers through sensing systems. The system also planned to use EPCglobal RFID tags to trace carbon footprint emissions at higher stages of the supply chain. However, as long as our knowledge, there has been no work that supports carbon credits by RFID technology.

9 Conclusion

The approach proposed in this paper can be proposed to solve serious problems with existing carbon credits, offsetting, and trading. The key idea underlying our approach is to introduce RFID tags (or barcodes) as physical certificates for the rights to claim carbon credits, including carbon emission credits. When purchasers buy items with credits for carbon offsets, they can claim the credits by returning the RFID tags (or barcodes) coupled with the credits to stakeholders, e.g., sellers or agencies, without the need for any complicated authentication. The approach can treat carbon credit trading as the trading of RFID tags. The approach was constructed to complement existing systems of supply chain management. It can be simply and intuitively provided in real supply chains. Our early experiment proved the feasibility and effectiveness of our approach. We believe the approach will be a powerful digital economy for carbon credits to reduce the emission of greenhouse gases on the earth.

References

1. Bang, M., Gustafsson, A., Katzeff, C.: Promoting new patterns in household energy consumption with pervasive learning games. In: de Kort, Y.A.W., IJsselsteijn, W.A., Midden, C., Eggen, B., Fogg, B.J. (eds.) PERSUASIVE 2007. LNCS, vol. 4744, pp. 55–63. Springer, Heidelberg (2007)
2. Dada, A., Staake, T., Fleisch, E.: The Potential of UbiComp Technologies to Determine the Carbon Footprints of Products. In: Proceedings of Pervasive Computing 2008 Workshop on Pervasive Persuasive Technology and Environmental Sustainability, pp. 50–53 (2008)
3. Froehlich, J., Dillahunt, T., Klasnja, P., Mankoff, J., Consolvo, S., Harrison, B., Landay, J.A.: UbiGreen: investigating a mobile tool for tracking and supporting green transportation habits. In: Proceedings of the 27th International Conference on Human Factors in Computing Systems (CHI 2009), pp. 1043–1052. ACM (2009)
4. Ilic, A., Staake, T., Fleisch, E.: Using Sensor Information to Reduce the Carbon Footprint of Perishable Goods. IEEE Pervasive Computing 8(1), 22–29 (2009)
5. IPCC fourth assessment report: Climate change. Technical report, IPCC (2007), `http://www.ipcc.ch/`
6. McCalley, T., Kaiser, F., Midden, C., Keser, M., Teunissen, M.: Persuasive Appliances: Goal Priming and Behavioral Response to Product-Integrated Energy Feedback. In: IJsselsteijn, W.A., de Kort, Y.A.W., Midden, C., Eggen, B., van den Hoven, E. (eds.) PERSUASIVE 2006. LNCS, vol. 3962, pp. 45–49. Springer, Heidelberg (2006)
7. Stern, N.: The Economics of Climate Change: The Stern Review. Cambridge University Press (2007)

Chromatic Scales on Our Eyes: How User Trust in a Website Can Be Altered by Color via Emotion

Jean-Eric Pelet, Christopher M. Conway, Panagiota Papadopoulou, and Moez Limayem

Abstract. Establishing customer trust on an e-commerce website possibly requires the provision of an environment in which customers can overcome their fear and reluctance about shopping transactions by forming trust and positive perceptions about the online vendor. This research concentrates on color as a central factor in

---

Jean-Eric Pelet
IDRAC International School of Management
University of Nantes
16 rue Laurence Savart
F - 75020 Paris
e-mail: jepelet@yahoo.com

Christopher M. Conway
IÉSEG School of Management
(LEM-CNRS)
1, parvis de la Défense
F - 92044 Paris-La Défense cedex
e-mail: c.conway@ieseg.fr

Panagiota Papadopoulou
University of Athens
Department of Informatics and Telecommunications
Panepistimiopolis, GR - 15784,
Athens, Greece
e-mail: peggy@di.uoa.gr

Moez Limayem
College of Business,
University of South Florida
Tampa, Florida
e-mail: mlimayem@usf.edu

P.-J. Benghozi, D. Krob, & F. Rowe (Eds.): *Digital Enterp. Des. & Manage. 2013*, AISC 205, pp. 111–121.
DOI: 10.1007/978-3-642-37317-6_10

the design of web pages that enhance users' positive emotional reactions as well as trust states and behaviors. Drawing on existing theories and empirical findings in the environmental psychology, human–computer interaction, as well as in the marketing and information systems research literature, a research model is developed to explain the relationships among background and foreground colors of a webpage, induced emotional responses in users, and users' trust toward the website as mediated by users' emotional states. A laboratory experiment was conducted to test the model and its associated hypotheses. This permitted assessment of the influence of background and foreground colors on user emotions and trust under varying brightness and saturation levels. Sixteen different graphic charts were evaluated based on various contrast ratios. We use the PAD (Pleasure Arousal Dominance) scale from Mehrabian and Russell (1974) and find that color can increase arousal, leading to stronger trust. Color leads to higher trust levels when arousal states are present at a high level. Our empirical findings provide valuable theoretical and practical implications regarding the effect of websites color on trust.

Keywords: Color, trust, trusting beliefs, e-commerce, arousal.

1 Introduction

Although e-commerce keeps on growing exponentially, designing interfaces still remains strategic for companies regarding their e-commerce website's ROI. Whether it comes from human–computer interaction (HCI), psychology and cognition, or from the marketing research, questions remain unanswered about the reason why consumers will trust a particular e-commerce website. As users have several options for shopping online, considering the huge number and variety of e-commerce websites, they may choose their preferred interface on the base of emotional responses. This places emphasis on the color of the webpage interface, and stresses the need for interfaces that promote engagement, pleasure, and delight rather than just functionality or ease-of-use (Marcus 2002; Wright *et al.,* 2001).

This paper aims to study the impact of the colors of e-commerce websites on customer trust. The effect of the colors of e-commerce websites on customer trust has only been indirectly examined in terms of website characteristics, such as website quality and usability, where color is viewed as a key interface attribute. In addition, despite the increased attention on affective aspects of user interface design (Dillon, 2001; Norman, 2003) and their importance for online shopping (Eroglu *et al.,* 2003), little is known about the emergence of trust thanks to greater experienced affective states. Affect is closely linked to attitudes, cognitions, and motivations. It influences and mediates specific aspects of interaction with a user interface (Deng and Poole, 2010).

The paper presents an empirical study of the effects of e-commerce website background and foreground colors on customer emotions and trust. Unlike most empirical studies dealing with color by comparing warm and cold colors, we examine color by focusing on its hue, brightness and saturation, following the recommendations of Gorn *et al.* (2004) and Pelet and Papadopoulou (2012) so as to demonstrate that its influence varies according to the intensity of each of these three components.

The structure of the paper is as follows. In the next section, the theoretical background and research model linking e-commerce, color, emotions and trust aspects are provided. The following section presents the research model and hypotheses. The empirical testing of the model is described next. The section that follows presents our results. The paper ends with the conclusions, implications for theory and practice, limitations and future research.

2 Theoretical Background and Research Model

The "Stimuli – Organism – Response" (SOR) environmental psychology model proposed by Mehrabian and Russell (1974), synthesized with the work of Engel *et al.* (1978) and Filser, (1994), is used as the framework in this study for understanding web users' emotional and trust responses to the foreground and background colors of a website. Our proposed model explains how the background and foreground colors of an e-commerce website and specifically their components - hue, brightness and saturation - can have an impact on the buyer's affective state of emotion and cognitive states of trust. Specifically, we use color as stimulus, the emotional state as arousal induced by the stimulus, and trust is the response.

Color has been identified as a pivotal component of e-commerce websites (Gorn *et al.*, 2004), supporting a positive customer experience. It contains three principal components (Trouvé, 1999):

- The hue (or chromatic tonality), which is the attribute of the visual sense defined according to the colors denominations such as blue, green, red;
- The saturation, which provides the proportion of chromatically pure color contained into the total sense;
- The brightness, which corresponds to the component according to which a surface illuminated by a source seems to emit more or less light.

To this day, the effects of the three color components on websites have been but seldom documented. Color has mainly been addressed in terms of warm and cold hues (Coursaris *et al.*, 2008; Papachristos *et al.*, 2005) and has not been examined with respect to its components, hue, brightness and saturation. Most of the studies linking color and e-commerce take into account balance or brightness as variables of colors (Brady and Phillips, 2003) or combinations of colors (Humar *et al.*, 2008), which do not allow for comparing the effects of the components of the colors. As Valdez and Mehrabian (1994), Drugeon-Lichtlé (2002), Camgöz *et al.* (2002), Gorn *et al.* (2004) and Pelet and Papadopoulou (2012) show about the brightness component of color, an experiment involving color should compare hue and brightness rather than warm and cold colors in order to understand its effects on consumer.

On a website, the interface represents a graphic chart, which refers to a collection of website elements. A graphic chart includes two colors, the foreground color and the background color, both of which constitute the color scheme. These colors create the contrast, which corresponds to the opposition between the foreground and the background colors, as defined by W3C (W3C, 2008). Its main function relies on facilitating the readability of the displayed information, and *a fortiori* the memorization process.

Aware of the significant and widely known impact of the atmosphere inside stores on the prospective buyers' activities and behavior in a traditional shopping context (Kotler, 1973; Donovan and Rossiter, 1982; Filser, 1994, 2003a, 2003b; Lemoine, 2003), there is a need to investigate how emotions and trust are affected by color, as an atmospheric variable and as a component of e-commerce interfaces. The role of color has been shown to be important for the readability and memorization of information, information seeking and navigation within a web vendor site (Pelet, 2008, 2010). These color-dependent factors constitute or are closely intertwined with antecedents of trust, which are associated with an e-commerce web site. This implies that color could be related to emotions, as it is a key attribute of website characteristics that have an effect on trust. Thus, we can hypothesize:

- H1A: An increase in contrast increases arousal
- H1B: An increase in foreground brightness increases arousal
- H1C: An increase in foreground saturation increases arousal
- H1D: An increase in background brightness decreases arousal
- H1E: An increase in background saturation decreases arousal:

However, color has only been implicitly associated with trust (see Pelet and Papadopoulou, 2011). The relationship between color itself and trust remains largely unaddressed. Color, as an important interface variable in e-commerce, is expected to be an important antecedent of customer trust in an online vendor.

There is also a lack of research on the effects of colors of e-commerce websites on consumer's trust, by considering the impact of affective states as a mediating variable of the link. The effect of color on emotions has been the research topic of several studies such as Valdez and Mehrabian (1994) and Lichtlé (2007) but not in an online context. Eroglu *et al.* (2001, 2003) investigated the effects of atmospheric cues of the online store on shoppers' emotional and cognitive states, showing that they affected their shopping outcomes. Similar to traditional in store stimuli, online colors can provide information about the retailer (e.g., the quality or type or retailer, the target audience of the retailer) as well as influencing shopper responses during the website visit (Eroglu *et al.*, 2003). The consumer's affective states can directly affect the website visit duration. It is then possible to presume that consumer's emotions affect the website visit trust of the consumer.

We develop the research hypotheses proposing links among the color, emotions and trust variables. Hence, we suggest the following hypotheses:

- H2A: An increase in pleasure increases trust
- H2B: An increase in arousal increases trust
- H2C: An increase in dominance decreases trust

3 Research Method

The research hypotheses have been empirically tested with a quantitative study in a laboratory experiment. This involved the use of a mock e-commerce site on which we manipulated the foreground and background colors. We used a single website for the experiment, so that our results would not be confounded by differences in the site. Consumers react well to tangible goods; therefore, the development and use of an e-commerce website selling CDs seemed appropriate, similar to the type of e-commerce website they might usually visit. In order to focus on only hue, brightness, and saturation, we decided to vary the site with only two hues, and two levels each of saturation and brightness. While this may reduce the strength of the effect observed, we felt that keeping the plan simple, in order to avoid confounds, was more important. After respondents interacted with the site, they filled out a questionnaire which we used to measure their emotional reactions and trust.

A laboratory experiment was necessary to allow for controlling and neutralizing the three major elements that should be taken into account when conducting a study focusing on color, namely the screens, the ambient light, and, the participants' color perception (Fernandez-Maloigne, 2004; Pelet and Papadopoulou, 2012). Since, these elements cannot be controlled in a distance study carried out over the Internet, a controlled laboratory setting had to be used for our experiment. Screens were calibrated using a probe to ensure that the colors were displayed as defined. The color of the walls was neutral grey. The grey color of the walls prevents a bad reflection of ambient light on the screens, which avoids color distortion on the screen. By neutralizing the colors of both the ambient light and the walls we were confident that we avoided any distortion. The brightness of the room was then set at 1000 lux to guarantee that the colored appearance of the websites used for the experiment would not be biased by a too dim or a too bright room light. On the website, foreground and background colors were controlled in order to provide the desire stimuli. Hues have been chosen in order to provide adequate contrasts for easy reading, following Hill and Scharff recommendations (1997). We made small changes to the saturation and brightness of both the foreground and background colors in the experiment. With low and high levels of both saturation and brightness, and both foreground and background colors, this resulted in 16 possible color combinations.

The laboratory experiment was conducted with 190 participants, recruited from undergraduate business classes and were offered course extra credit for participation. A respondent would visit the website and browse the available CDs. There were 57 available CDs in 19 categories (3 CDs/category). For each CD, participants could see the CD cover, the album title, the artist name, and seven information items: music category, online store price, music company price, sale percentage, delivery time, state (new or used) and delivery charge. This detailed information was displayed by clicking on the cover image or the title of a CD. In addition, the detailed information included a short description of around 20 words, next to the CD cover.

Participants could select a category on the left side of the webpage and see the 3 CDs of this category on the right side of the same webpage. They had to look into the details of a minimum of two CDs of their choice. They could look at more than two CDs if they wanted to, and add them to their shopping cart, but they could not conduct real purchases.

Each participant visited the website with colors which were randomly selected among the 16 graphic charts prepared for the experiment, as explained previously. A balanced distribution of the graphic charts among all respondents was ensured. After viewing at least two CDs, an easy to see link appeared and respondents were asked to complete a questionnaire with questions about emotions and trust. We used previously validated measures for trust and affective responses. The trust measures were based on the Pavlou and Gefen (2004) and McKnight *et al.* (2002) scales. Emotions were measured with the Mehrabian and Russell Pleasure, Arousal, and Dominance (PAD) scale. Participants were not able to proceed to the questionnaire unless they had visited at least two CDs, in order to ensure that they had viewed adequate information for responding to the subsequent questions. Then, each participant was asked to go to another room to take an Ishihara test, the most common test for detection of color blindness. It was conducted in a different room from the experiment, as suggested by Lanthony (2005) to ensure that participants were in a position to respond to the stimulus. It also prevented hypothesis guessing.

4 Analysis and Results

We collected the survey responses by including pages in the mock website in standard fonts and colors without ornamentation. After discarding questionnaires that were incomplete or filled by color blind respondents, 190 valid responses were used for the analysis, with each color scheme being visited by about 12 respondents. There were 16 color deficient participants, which account for approximately 8% of the sample. This percentage is equal to the actual percentage of color deficient males in the world (Brémond, 2002).

Because respondents do not always respond carefully to studies, we cleaned the data prior to analysis. We imported the data into R version 2.15.1 (R Development Core Team, 2011) for this process. We cleaned the data by visually comparing the scatter plot of the Mahalanobis distance versus a chi-square distribution using a diagonal as a visual referent, removing extreme data points incrementally until the distributions matched. Most deleted points were invalid responses. While a few may have been valid, though extreme, responses, we were interested in finding the average person's response, so deleting extreme points is justified as well. After cleaning, 138 data points remained.

We loaded the cleaned data into SmartPLS version 2.0.3 for analysis (Ringle, Wende and Will, 2005). We used a first-order model, using the default PLS algorithm options, to obtain the latent variable scores for the trust variable. The measures were reliable, with composite reliability ranging from 0.83 (integrity) to 0.96 (competence). Average variance extracted (AVE) ranged from 0.46 (integrity) to 0.77 (competence). The AVE for integrity was a little below the threshold of 0.50. An examination of the item loadings indicated that there was some cross loading between benevolence and integrity. This is not a serious problem for our results, particularly since they were going to be loaded onto the same second-order factor.

We constructed the final model with trust as a second-order factor, using the default PLS options, and ran a bootstrap with 138 cases for 1000 iterations to estimate the standard errors. The results are presented in Table 1 and Figure 1. The AVE for arousal was a little low; however, we judged this not to be a serious problem, since reliability is good, and the discriminant validity is still good. Since the color components and contrast are single-indicator values, and trust is a formative construct, there are no reliability figures associated with them.

Table 1 Convergent and discriminant validity

Variable	**AVE**	**Composite Reliability**	**Correlations (diagonal is sqart (AVE))**			
			Pleasure	**Arousal**	**Dominance**	**Trust**
Pleasure	0.991	0.998	0.995			
Arousal	0.470	0.830	0.332	0.686		
Dominance	0.767	0.950	-0.407	-0.051	0.876	
Trust	Not applicable		0.603	0.367	-0.387	Not Applicable

Our results support the hypotheses, as seen in Figure 5. Contrast, foreground saturation, and foreground brightness have a significant positive relationship with arousal; higher values of these variables lead to higher arousal, supporting H1A, H1B and H1CHigher values of background saturation reduce arousal supporting H1E. However, there is no significant relationship between background brightness and arousal, thus H1D is not supported.

It is interesting to note that there are significant effects of both contrast and color components independent of each other. In models which do not include contrast, no color components are significant; and, likewise, if the color components are not included, then contrast is not significant. Thus, there is a suppression effect, in which not including both types of variables yields inconclusive results.

Increase in pleasure or arousal result in an increase in trust, confirming H2A and H2B. A person who feels more pleased or aroused when navigating a website

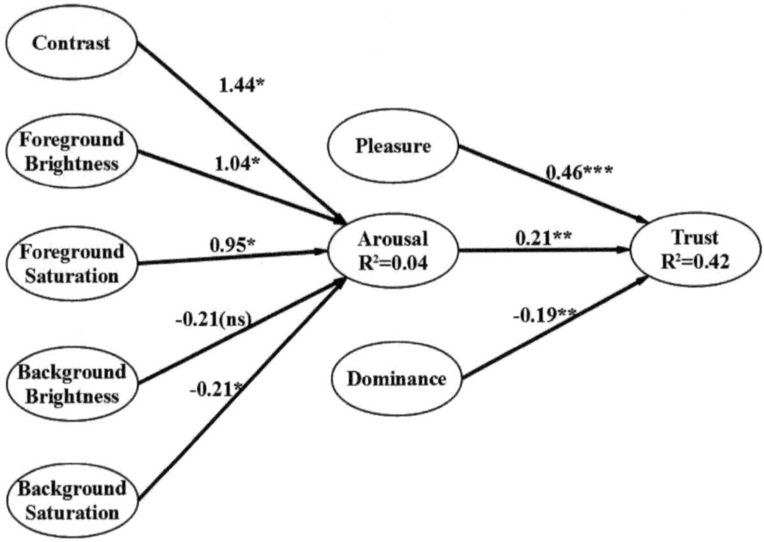

Fig. 1 Results – (*p<0.05, **p<0.01, ***p<0.001)

is more willing to trust it. An increase in dominance results in a decrease of trust, confirming H2C. An increase in dominance means that the person feels more in control of the relationship between themselves and the website. In these circumstances, the person feels less need to trust the object of their attention, because of that feeling of control. With less need to trust, the person will not expend cognitive effort in determining trust, leading to a lower overall level of trust.

5 Discussion and Conclusion

Traditional e-commerce websites use a similar, almost sterile, design solution, with muted colors and standard placement of navigational aids. For example, Dell, Microsoft, Facebook and so, on use almost interchangeable designs, making us more accustomed to these functional, stereotypical websites. Our research indicates that this approach may not be optimal; that rather than not upsetting users, web designers and organizations might be more successful if they aim for inspiring users. Among the characteristics of an e-commerce website, colors have a significant influence on the reaction of the consumers. These colors can make them comfortable, but also uninspired, and their response is correspondingly weak. Our research indicates that using more saturated and brighter foreground colors can motivate consumers to feel excited or energized by a website. Naturally occurring color combinations with bright, saturated foreground colors that are reminiscent of natural environments lead the consumer to more positive reactions to the site. These kinds of colors make the site more memorable and more pleasant for the consumer to use. These are all positive emotions, and consumers naturally try to justify their positive emotions by attributing more positive characteristics to a website which induces these emotions. Our research further shows that arousal is an important component of those reactions, which can lead to in increase in trust. Increasing trust is an important goal of website design, as trustworthiness is well known to improve the success of an e-commerce website.

Obviously, the color is only one component of the characteristics that lead a consumer to react positively to and trust a website. Other elements which affect the atmospherics of a site, such as familiarity *("the navigation bars and important cues of the website are where I expect them to be")*, clarity of interface *("the controls do what I expect them to do and do not increase my cognitive load")*, and transparency of information use *("I have assurances that my data privacy will be respected and my data will be used to my benefit")* are also critical (Eroglu *et al.* 2003). When an organization properly uses all these components, including color, it can create a truly memorable and satisfying experience for the consumer. Further research will help to identify which of these components will stand with color as determinants of those positive consumer reactions.

References

Bellizzi, J.A., Hite, R.E.: Environmental Color, Consumer Feelings, and Purchase Likelihood. Psychology and Marketing 9, 347–364 (1992)

Brady, L., Phillips, C.: Aesthetics and usability: a look at colour and balance. Usability News 5(2) (2003), `http://www.surl.org/usabilitynews/112/pdf/Usability% 20News%20112%20-%20Phillips.pdf`

Brémond, É.: L'intelligence de la couleur. Albin Michel, Paris (2002)

Brengman, M., Geuens, M.: The Four Dimensional Impact of Color on Shopper's Emotions. Advances in Consumer Research 31(1), 122–128 (2004)

Camgöz, N., Yener, N., Güvenç, D.: Effects of hue, saturation, and brightness on preference. Color Research & Application 27(3), 199–207 (2002)

Deng, L., Poole, M.S.: Impacts of Web Page Visual Complexity and Order. MIS Quarterly 34(4), 711–730 (2010)

Dayal, S., Landesberg, H., Zeisser, M.: Building trust online. McKinsey Quarterly (October (2001), `http://www.mckinseyquarterly.com/ab_g.asp?ar=1138`

Dillon, A.: Beyond Usability: Process, Outcome and Affect in Human–Computer Interactions. PANTIN-SOHIER G (2004); Le rôle de la couleur dans la perception des traits de personnalité de la marque: une étude comparative Etats-Unis / Suède, 1ère journée de recherche AFM-AUDENCIA sur le Marketing et le Design - Janvier 26(4), 57–69 (2001)

Dunn, B.: Choice of Color for Product Can Be Critical Factor, The Gazette, August 10, 6 (1992)

eMarketer US Consumers Wary of Online Payments (2008), `http://www.emarketer.com/SiteSearch.aspx?arg=trust&src=search_go_welcome`

Engel, J.F., Kollat, D.T., Blackwell, R.D.: Consumer Behavior, 3rd edn., vol. II. Dryden Press, Hinsdale (1978)

Eroglu, S.A., Machleit, K.A., Davis, L.M.: Atmospheric qualities of online retailing: a conceptual model and implications. Journal of Business Research, special issue on Retail Strategy and Consumer Decision Research 54, 177–184 (2001)

Eroglu, S.A., Machleit, K.A., Davis, L.M.: Empirical Testing of a Model of Online Store Atmospherics and Shopper Responses. Psychology & Marketing 20(2), 139–150 (2003)

Fernandez-Maloigne, C.: Quelle métrique pour l'évaluation de la qualité couleur d'une image numérique? In: Application à la compression JPEG 2000, CAM Conférence, Paris (October 2004)

Filser M.: Le comportement du consommateur, Précis Dalloz, Paris (1994)

Gefen, D., Karahanna, E., Straub, D.W.: Trust and TAM in Online Shopping: An Integrated Model. MIS Quarterly 27(1), 51–90 (2003)

Gorn, G., Chattopadhyay, A., Sengupta, J., Tripathi, S.: Waiting for the web: how screen color affects time perception. Journal of Marketing Research XLI, 215–225 (2004)

Hill, A., Scharff, L.V.: Readability of websites with various foreground/background color combinations, font types and word styles. In: Proceedings of 11th National Conference in Undergraduate Research, vol. 2, pp. 742–746 (1997)

Howard Hughes Medical Institute, Breaking the Code of Color: Color Blindness: More Prevalent Among Males, Retrieved from the Internet at (2006), `http://www.hhmi.org/senses/b130.htmlthe25/09/2012`

Humar, I., Gradisar, M., Turk, T.: The Impact of Colors on the Speed of Reading Web Pages on a CRT monitor. International Journal of Industrial Ergonomics 38, 885–899 (2008)

Lichtlé, M.-C.: The effect of an advertisement's colour on emotions evoked by an ad and attitude towards the ad The moderating role of the optimal stimulation level. International Journal of Advertising 26(1), 37–62 (2007)

Marcus, A.: The Cult of Cute: The Challenge of User Experience Design. Interactions 9(6), 29–34 (2002)

McKnight, D.H., Choudhury, V., Kacmar, C.: The Impact of Initial Consumer Trust on Intentions to Transact with a Web Site: A Trust Building Model. Journal of Strategic Information Systems 11, 297–323 (2002)

Mehrabian, A., Russell, J.A.: An Approach to Environmental Psychology. MIT Press (1974)

Munsell, A.: The Munsell Colour Atlas. Munsell Color Corp (1969)

Norman, D.A.: Emotional Design: Why We Love (Or Hate) Everyday Things. Basic Books, New York (2003)

Pantin-Sohier, G.: Le rôle de la couleur dans la perception des traits de personnalité de la marque : une étude comparative Etats-Unis / Suède, 1ère journée de recherche AFM-AUDENCIA sur le Marketing et le Design Janvier (2004)

Papachristos, E., Tselios, N., Avouris, N.: Inferring relations between color and emotional dimensions of a Web site using Bayesian Networks. In: Proceedings of Interact 2005 (2005)

Pavlou, P.A., Gefen, D.: Building Effective Online Marketplaces with Institution-Based Trust. Information Systems Research 15(1), 37–59 (2004)

Pelet, J.-É., Papadopoulou, P., et al.: The effect of colors of e-commerce websites on consumer mood, memorization and buying intention. European Journal in Information Systems (EJIS), Special Issue: Information Research, Education and Policy in the Mediterranean Region (2012)

Pelet, J.-É., Papadopoulou, P., et al.: The Effect of E-commerce Websites' Colors on Customer Trust. International Journal of E-Business Research 7(3), 1–18 (2011)

R Development Core Team, R: A Language and Environment for Statistical Computing, Vienna, Austria: R Foundation for Statistical Computing (2011)

Ringle, C.M., Wende, S., Will, A.: SmartPLS 2.0 M3 (beta), University of Hamburg, Hamburg (2005)

Trouvé, A.: La mesure de la couleur, Paris, CETIM, éd. Association française de normalisation (AFNOR) (1999)

Valdez, P., Mehrabian, J.: Effects of color on emotions. Journal of Experimental Psychology, General 123(4), 394–409 (1994)

W3C Publication Version 1.1 of the AccessiWeb repository created by the W3C, of the June 9, 2008 (2008), http://www.w3c.org/

Wright, P., McCarthy, J., Marsh, T.: From Usability to User Experience. British HCI Group Members Magazine (46), 4–11 (2001)

Can Agile Collaboration Practices Enhance Knowledge Creation between Cross-Functional Teams?

Carine Khalil, Valérie Fernandez, and Thomas Houy

Abstract. Agile philosophy emphasizes constant interactions and close collaboration between team members. This emergent management philosophy relies on a set of practices that aim at creating an environment in which teams are able to respond rapidly to customer's needs and to deal effectively with changing situations. From this perspective, agile practices can be viewed as a way to enhance knowledge creation and knowledge sharing between team members. However these emergent practices necessitate an environment that facilitates communication and coordination mechanisms. The present paper aims at analyzing how large organizations, characterized by distributed and cross-functional teams, can cultivate an agile environment where inter-individual knowledge exchanges are encouraged. Even though mutual adjustments and face to face interactions are not easily achieved in large and distributed organizations, the contributions of agile practices in such contexts remain significant. These practices can foster knowledge development and collective learning processes and subsequently improve organization's adaptability.

Keywords: Agile Methods, Cross-Functional Teams, Learning Organization, Collaboration Practices.

Carine Khalil
Ecole Centrale, Grande Voie des Vignes,
92295 Châtenay-Malabry, France
e-mail: carine.khalil@ecp.fr

Valérie Fernandez · Thomas Houy
Télécom ParisTech, 46 rue Barrault,
75013, Paris, France
e-mail: {valerie.fernandez,thomas.houy}@telecom-paristech.fr

P.-J. Benghozi, D. Krob, & F. Rowe (Eds.): *Digital Enterp. Des. & Manage. 2013*, AISC 205, pp. 123–133.
DOI: 10.1007/978-3-642-37317-6_11

1 Introduction

In dynamic and competitive environments, "traditional" software methodologies, based on comprehensive planning, detailed documentation and design have been progressively questioned by a number of practitioners. These methods are considered as unable to deal with rapid technological innovations and changing demands. In this respect, a new style of software methodology called agile methods has emerged and gained popularity within software industries. Literally, the "agile" concept refers to "having a quick resourceful and adaptable character". In software development, agile methods consist on delivering, at the end of each iterative cycle[1], a functional version of the product. This iterative mode enhances feedback and adaptation to continuous changes and seems to better satisfy customers' demands by delivering a concrete version of the product with the required set of functionalities.

Different empirical studies have been conducted lately assessing the development methods used by agile practitioners. These surveys highlight the growing adoption level of agile practices and tools (Ambler[2], 2008; Version One[3], 2010; Enquête Nationale[4], 2011). These emerging management and development methods seem to improve time-to-market, team productivity and product quality.

Agile methods rely on a set of practices and tools that aim at creating an environment in which development teams are able to respond rapidly to customer's demands and to deal effectively with changing situations. Therefore, agile development advocates constant interactions and exchanges between team members. It is an attempt to bring people together and to enhance collective practices and achievements. From this perspective, agile practices can contribute to knowledge development. They convert new knowledge through shared experiences. However, these collaboration practices are more adequate to organizational contexts where communication and coordination mechanisms can be easily achieved. In other words, in a workplace where team members are collocated, spend time together, often discuss and interact with each other. What about large organizations characterized by transversal project teams? How can agile practices foster knowledge creation processes in such contexts?

Few empirical studies have focused on how agile practices embrace knowledge development in large organizations. We adopted an ethnographic approach in order to address these relevant questions and better understand the context in

---

[1] The main idea of an iterative approach is to divide the development cycle into simpler and more manageable units. Each unit is analyzed, designed and implemented in order to produce an executable deliverable with a limited set of functionalities. The final executable deliverable encompasses all the functionalities expected from the system.

[2] `http://www.ambysoft.com/surveys/agileFebruary2008.html`

[3] `http://www.versionone.com/pdf/2010_State_of_Agile_Development_Survey_Results.pdf`

[4] `http://fr.slideshare.net/xwarzee/enqute-2011-vous-votre-organisation-et-agile`

which the study was taking place. Accordingly, we conducted a set of semi-structured interviews with cross-functional team members that are using agile collaboration practices. A general inductive approach has been adopted for analyzing and interpreting the collected data.

In the following section (2), we will highlight the role of knowledge creation in agile environments and we will review the work related to scrum practices. In section (3), we will present the research methodology and the context of the study. We will portray the results of the research work in section (4). And finally, we will discuss the findings and the limitations of the study in section (5).

2 Knowledge Creation in Agile Environments

In today's business environment, software industries require a flexible organization that allows project teams to adapt quickly to inevitable changing demands and requirements. In this respect, software practitioners have suggested a set of relatively new management practices that address the challenges faced in unpredictable environments where reactivity and adaptability become fundamental. These practices, called "agile" practices, stress on close collaboration and frequent feedback between team members and the client. They encourage experience and knowledge sharing and subsequently increase the capabilities of team members to cope with uncertain and ambiguous situations.

According to [1], organizational knowledge is created through the continuous social interaction of tacit and explicit knowledge. While explicit knowledge can be codified and transferred through formal communication or documentation (theoretical approaches, problem-solving, manuals and databases), tacit knowledge is more difficult to transfer. It involves both cognitive (mental models of the world) and technical elements (concrete know-how and skills). Dialogue is an important means for elicitating and translating the tacit knowledge into a readily understandable form [2]. Agile methods denote a social approach where groups and individuals constantly interact with each other, collectively construct meaning of the ongoing flow of experience and act accordingly. These methods rely on a set of collaboration practices that promote inter-individual exchanges and enhance knowledge creation and sharing. Thus, agile organizations can thrive and respond better to customer's demands [3-4]. The following section introduces rapidly the scrum method and stresses on how this method helps organizations to achieve sustainable competitive advantages.

2.1 Scrum: Principles, Practices and Management Tools

Different agile methods exist. In the present work, we will focus on scrum method. Scrum is an iterative and incremental approach for managing projects. Three pillars sustain the development process: transparency, inspection and adaptation [5]. The main purpose of scrum is to foster team productivity by providing "light" management practices (pre-sprint, daily sprint, retrospective meetings, product backlog, burndown charts, etc.) and creating an environment where teams can easily communicate and adapt to changes. Through continuous feedback and interactions, scrum provides a context where project teams are predisposed to combining and creating knowledge. Hence, communication is considered a critical factor for scrum teams. Table 1 summarizes the characteristics of scrum.

Table 1 Components[5] of Scrum method

Scrum	
Principles	Transparency, Inspection and Adaptation.
Roles	Scrum Master, Product-Owner, Scrum Team
Management Practices	Pre-sprint, Sprint Planning Meeting, Sprint, Post-Sprint Meeting and Retrospective Meeting.
Management Tools	Product Backlog, Sprint Backlog, Burndown charts

2.2 Collaboration Practices with Scrum: The Related Work

Different empirical studies have examined the impact of scrum on team collaboration. In this research paper, we stress on three scrum management and collaboration practices in order to explore their impact on knowledge creation: daily scrum, iterative development and scrum whiteboard. Daily meetings such as "daily scrums" strengthen the communication between team members [6-7], improve information sharing [8] and collective problems solving [9-10]. They clarify the status of the on-going operations and sheds light on the difficulties encountered throughout the project. It also ensures a better control of the project and provides a coordination mechanism for everyone in the project [11]. However, in distributed environments, frequent and informal communication is hard to achieve impacting the collaboration between team members [12] and the pursuit of a common goal [13]. Furthermore, daily scrums are difficult to realize across long distances and geodistributed teams [14-16]. In such environments, team members should be equipped with different communication medias and information technologies in order to facilitate their direct communication and documents sharing [17-18]. In addition to the previous collaboration practice, the iterative development encourages collaboration between team members and the client [6], [14]. This practice adds agility to the development process by providing rapid feedback on the implemented functionalities [19]. It facilitates the monitoring of the project progress [15] and improves organizational learning [20] by incorporating feedback into future iterations [8]. Some practitioners stated that frequent iterations combined with client voice can lead to successful results [20]. However, it is hard to manage iterations that run in parallel, on different geographical sites [19], [21]. Least but not last, the whiteboard plays an important role in the organizational learning processes. The use of this tool enables team members to have a sharing vision of the project requirements [22- 23] and create an informative workspace [9-10].

[5] Scrum practices, tools and roles are defined in Annex I.

However, physical distance across teams constrains the information exchanges. These empirical studies stress on how agile practices promote feedback and collaboration, encourage information exchanges between team members and improve collective problems-solving. An agile organization is therefore considered as a continuous flow of organizing and learning processes where frequent communication and collaboration are fundamental. However, these practices are not easily implemented in large organizations characterized by cross-functional and geodistributed teams. Up till now, few studies have investigated the way agile practices can develop organizational learning in distributed environments. Our research work aims at understanding, from cross-functional teams' perspective, how agile practices can foster the mutual development and the sharing of tacit knowledge within their organization.

3 The Research Context

3.1 *Context of the Study*

The case study was carried out in a division of a French telecommunication company.

The organizational structure of the studied entity can be categorized as a lightweight[6] structure [24] characterized by distributed teams. The project manager coordinates the activities of his team members and facilitates the information exchange between parts of the organization. Projects are generally large, involving approximately thirty five persons each. They combine cross-functional actors that are hierarchically attached to different functional managers and they intervene temporarily on different projects in parallel. This French entity operates in a dynamic and competitive technological environment necessitating highly adaptive project management systems. In such a turbulent context, the top management decided to implement new management practices in order to improve the team cohesion in addition to the project transparency. The lack of communication between cross-functional teams increased the need for tighter collaboration. In this respect, top management decided to implement, within the project teams, a set of scrum practices that emphasize communication and collaboration. This includes daily meetings, virtual whiteboards and kaizen[7] sessions.

---

[6] The lightweight team structure is an organizational form typically found in large, mature firms. The team members physically reside in their functional areas where each functional organization has a representative in the project team. The project manager, who is typically chosen out of the function that is most vested in the development process, is responsible for coordinating the activities of the different functions. However, the project manager has little authority and influence. The project team members remain under the control of their respective functional managers.

[7] It's a Japanese term that means continuous improvement. Kaizen events consist of gathering operators, managers, owner of a process in one place, mapping the existing process in order to improve it.

3.2 Methodology

We adopted a qualitative approach based on an instrumental case study [25]. We focused on a single case study to investigate, in-depth, the context settings (physical, organizational and technical conditions) in which project teams operate. 15 interviews were conducted with project team members. The interviewees were selected based on their will to participate in the study. The interviews were semi-structured and each lasted one hour. The interviews were recorded with the approval of the interviewees who were explicitly informed about the purpose of the study. These interviews aimed at identifying how project teams perceive the implementation of agile practices within their organization and how such collaboration practices influence knowledge creation and capitalization between the teams. We also carried out informal discussions with the participants throughout the study period. Complementary data such as e-mails and documents were also collected to enhance our understanding of the context. For data analysis, we adopted an interpretive approach. We began with multiple readings of our field notes to better understand the context in which the project was taking place. The research question "How can agile practices foster knowledge creation processes in such contexts?" has guided us in identifying the key concepts in each sentence or/and paragraph. These meaningful segments were classified into categories, where, each refers to a particular meaning. A set of inductive categories were subsequently defined and justified with verbatim. Among these, we cite the following: management issues, organizational structure, team composition, inter-individual interactions.

4 Results

The following section describes the way the interviewees make sense of agile practices and their impact on the software development process. Even though project team members are aware of the benefits of these collaboration practices, they express scepticism towards their implementation within large organizations and distributed teams. A set of contextual factors has been identified as barriers to knowledge sharing and development.

The geographical distribution of the project teams is considered to be a challenge for gathering the whole team and fostering experiences exchanges between team members. Thus, the lack of face-to-face interactions constrains tacit knowledge capitalization. Even though information and communication technologies enable real-time communication and document exchanges, they cannot replace direct contact where tacit knowledge can be converted to explicit knowledge and transmitted. In addition, the creation of a virtual information environment is very demanding. Storyboards must be often updated and controlled in order to enable smooth coordination between distributed teams.

"*We are geographically distributed and there are a lot of cross-functional animations... It's not possible to share our daily experiences if we are distributed geographically... especially if we meet once a week. It doesn't promote close collaboration*" (project manager); "*Information and communication technologies can help us share documents and communicate over distance but it's different when the team is colocated....The geodistribution can skew the information*" (product development manager).

Large scale projects are also viewed as a challenge for implementing collaboration practices. The implementation of daily meetings that encompasses all the project team members was challenging for the project managers. The project managers aim to reduce the number of participants in order to respect the fifteen minutes time-boxed meeting. Consequently problem sharing and capitalization could not be done properly due to the non-participation of some key members. The interviewees were not encouraged to implement additional meetings. The big number of existing meetings and their long duration discourage the participants to attend supplementary ones. Moreover, kaizen sessions necessitate the involvement of the whole team to make them successful. According to the interviewees these sessions are not efficient if they are done in an isolated way.

"*Managing large teams is so challenging... I can't see how I can include all the project team in one meeting... it's not possible unless the meeting lasts for several hours... In small colocated teams, actors can directly deal with their neighbours if any problem occurs ... In large projects it's different*" (project manager); "*The way we run kaizen sessions cannot optimize the continuous improvement... it is absolutely necessary to involve the entire team*" (project manager).

Another contextual factor we identified is the team's composition. It seems to influence the knowledge development and transfer between project team members. The involvement of project members in different projects simultaneously constrains frequent exchanges and knowledge creation. The lack of time resources disables their participation in collective activities that promote knowledge sharing. Teams couldn't attend the daily meetings organized by the project manager. Furthermore, the updating of the virtual storyboard and its sharing between the teams were also difficult. The creation of a common and a well structured database requires a constant control that guarantees high data quality. Yet, in the studied context, the common database was not well managed and organized. There were missing reports and documents.

"*It is difficult to promote knowledge sharing since each team manages its own planning .... There is a movement within the team... people intervene at some point and then they move out which leads to a loss of information...* " (project manager); "*we have some teams that externalize some of their work which decrease the project visibility*" (project manager); "*the existing database is not reliable... We don't have a system that verifies the data entry*" (project manager).

In addition to the cited contextual factors, the lack of authority of the project manager was also perceived as a challenge for encouraging collective learning processes. The coordination of different functional teams has limited the circulation of instruction and information. It was difficult to foster collaboration and interactions between different functional teams.

"The role of the project manager is limited to an orchestra leader ...we don't control the activities of our project teams...each one of them has its own constraints and priorities" (project manager).

Thus, the implementation of agile practices within transverse teams and geodistributed environment is challenging. Agile practices necessitate a structuring context, frequent communication and continuous feedback. Furthermore, these practices necessitate an organization where project teams work under the authority of a project manager. The studied actors highlighted their preoccupations regarding the context in which they operate. The project size, the organizational structure and the team composition seem to influence the efficiency of agile practices and their impact on collective learning processes. The perceived usefulness of agile practices is not sufficient to successfully integrate these practices within project teams. The creation of a collaborative learning environment is very demanding when the teams are distributed and involved in different projects at the same time.

Our interpreted data underlined the challenges faced by actors in a lightweight organizational structure. In this respect, we can imagine a reorganization of the studied context that optimizes the use of agile tools and practices. All key members must participate to daily meetings and kaizen sessions in order to ensure knowledge capitalization and collective problem-solving.

5 Conclusion

The field notes have shown that the organizational context can be challenging while implementing agile practices and tools. Therefore, we believe that beyond the contextual factors, it seems fundamental to integrate agile methods as structured learning approaches. The organizational agility can be achieved through the capability of its members to rapidly reconfigure their resources and adapt to changes. Agile practices can be viewed as contributors to knowledge development. By integrating these tools in their daily work, development teams become more competent, able to respond to ambiguous situations and subsequently organizations become more agile and develop a customer responsive culture. This paper highlights obstacles faced by cross-functional teams working in a geodistributed environment and stresses on the need for thriving towards a learning organizations by adapting and integrating properly agile tools and practices. Nevertheless, this study presents two major limitations. First, the research results are limited to a single study constraining their generalization to other contexts. Hence, its application to other contexts and teams can constitute a further step of the study. Furthermore, this study has treated a limited number of agile practices. In the future, it would be interesting to consider more agile collaboration and engineering practices.

References

1. Nonaka, I., Takeuchi, H.: The Knowledge-Creating Company. Oxford University Press, New York (1995)
2. Nonaka, I.: A dynamic theory of organizational knowledge creation. Organization Science 5(1), 14–37 (1994)
3. Beck, K.: Extreme Programming Explained: Embrace Change. Addison-Wesley Professional (1999)
4. Poppendieck, M., Poppendieck, T.: Implementing lean software development: from concept to cash. Addison-Wesley Professional (2006)
5. Schwaber, K.: Agile project management with Scrum. Microsoft Press (2004)
6. Svensson, H., Höst, M.: Views from an organization on how agile development affects its collaboration with a software development team. In: Bomarius, F., Komi-Sirviö, S. (eds.) PROFES 2005. LNCS, vol. 3547, pp. 487–501. Springer, Heidelberg (2005)
7. Chong, J.: Social behaviors on XP and non XP: A comparative Study. In: The Agile Development Conference, pp. 39–48. IEEE Computer Society (2005)
8. Melnik, G., Maurer, F.: Perceptions of agile practices: A student survey. In: 2nd XP Universe and First Agile Universe Conference on Extreme Programming and Agile Methods, pp. 241–250 (2002)
9. Robinson, H., Sharp, H.: The characteristics of XP teams. In: Eckstein, J., Baumeister, H. (eds.) XP 2004. LNCS, vol. 3092, pp. 139–147. Springer, Heidelberg (2004)
10. Sharp, H., Robinson, H.: Collaboration and Coordination in mature Extreme Programming teams. International Journal of Human Computer Studies 66(7), 506–518 (2008)
11. Paasivaara, M., Durasiewicz, S., Lassenius, C.: Using Scrum in Distributed agile development: A multiple case study. In: 4th International Conference on Global Software Engineering, pp. 195–204. IEEE (2009)
12. Simons, M.: Internationally Agile. InformIT (2002), `http://www.informit.com/articles/article.aspx?p=25929`
13. Paasivaara, M., Durasiewicz, S., Lassenius, C.: Distributed agile development: Using scrum in a large project. In: 3rd International Conference on Global Software Engineering, pp. 87–95. IEEE (2008)
14. Highsmith, J., Cockburn, A.: Agile software development: the business of Innovation. IEEE Computer 34(9), 120–122 (2001)
15. Begel, A., Nagappan, N.: Usage and perceptions of agile software development in an Industrial Context: An Exploratory Study. In: 1st International Symposium on Empirical Software Engineering and Measurement, pp. 255–264. IEEE Computer Society (2007)
16. Yap, M.: Follow the sun: distributed extreme programming development. In: Agile Conference, pp. 218–224 (2005)
17. Jensen, B., Zilmer, A.: Cross-continent development using Scrum and XP. In: 4th International Conference on Extreme Programming and Agile Processes in Software Engineering, pp. 146–153 (2003)
18. Braithwaite, K., Joyce, T.: XP expanded: distributed extreme programming. In: 6thInternational Conference on Extreme Programming and Agile Processes in Software Engineering, pp. 180–188 (2003)
19. Karlström, D., Runeson, P.: Combining agile methods with stage-gate project management. Software IEEE 22(3), 43–49 (2005)

20. Middleton, P., Flaxel, A., Cookson, A.: Lean software management case study: Timberline Inc. In: 6th International Conference on Extreme Programming and Agile Processes in Software Engineering, pp. 1–9 (2005)
21. Sutherland, J., Viktorov, A., Blount, J., Puntikov, N.: Distributed scrum: agile project management with outsourced development teams. In: 40th Annual Hawaii International Conference on System Sciences, p. 274 (2007)
22. Danait, A.: Agile offshore techniques-a case study. In: Agile Development Conference, pp. 214–217 (2005)
23. Sharp, H., Robinson, H., Petre, M.: The role of physical artifacts in agile software development: Two complementary perspectives. Interacting with Computers 21(1-2), 108–116 (2009)
24. Wheelwright, S.C., Clark, K.B.: Revolutionizing Product Development: Quantum Leaps in Speed, Efficiency and Quality. Free Press, New York (1992)
25. Stake, R.E.: The art of case study research. Sage, Thousand Oaks (1995)

Appendix I – Glossary of Agile Terms

Burndown Chart: It shows work remaining over time. Work remaining is the Y axis and time is the X axis. The work remaining should jig up and down and eventually trend downward.

Daily Scrum: It's a fifteen-minute daily meeting for each team member to answer three questions: what have I done since the last scrum meeting? What will I do before the next scrum meeting? And what prevents me from performing my work as efficiently as possible?

Kaizen: it's a Japanese term that means continuous improvement. Kaizen events consist on gathering operators, managers, owner of a process in one place, mapping the existing process in order to improve it.

On-site customer: It consists on having a real, live user that constantly collaborates with the development team. The on-site customer is available full-time to answer questions.

Post-sprint Meeting: At the end of the sprint iteration, a post-sprint meeting is held to review progress, demonstrate features to the customers and review the project from a technical perspective.

Product Backlog: The product backlog is the requirements for a system, expressed as a prioritized list of product backlog items. These included both functional and non-functional customer requirements, as well as technical team-generated requirements. While there are multiple inputs to the product backlog, it is the sole responsibility of the product owner to prioritize the product backlog.

Product Backlog item: In Scrum, a product backlog item ("PBI", "backlog item", or "item") is a unit of work small enough to be completed by a team in one Sprint iteration. Backlog items are decomposed into one or more tasks listed in a sprint backlog.

Product-Owner: In Scrum, a single person must have final authority representing the customer's interest in backlog prioritization and requirements questions. This person must be available at any time especially during the sprint planning meeting and the sprint review meeting.

Retrospective meeting: The sprint retrospective meeting is held at the end of every sprint after the sprint review meeting. The team and Scrum-Master meet to discuss what went well and what to improve in the next sprint.

Scrum-Master: The Scrum-Master is a facilitator for the team and product owner. Rather than managing the team, the Scrum-Master works to assist both the team and product owner.

Scrum team: It consists of seven plus or minus two people. For software development projects, the team members are usually a mix of software engineers, architects, programmers, analysts, QA experts, testers, UI designers, etc.

Sprint: It defines the work for a sprint, represented by the set of tasks that must be completed to realize the sprint's goals, and the selected set of product backlog item.

Sprint Planning Meeting: The Sprint planning meeting is a negotiation between the team and the product owner about what the team will do during the next sprint. The product owner and all team members agree on a set of sprint goals, which is used to determine which product backlog items will be implemented in the next sprint. Then, the Scrum-Master and his team focus on how the selected product items will be implemented. .

Stand-up meeting: It’s a fifteen daily meeting for XP teams. During this meeting, developers share their experiences of the day before, talk about their progress since the last stand-up and the anticipated work until the next stand-up.

Story-cards: They represent brief details of the tasks being actively worked upon.

[illegible] In Scrum, a single person must have final authority representing the customer's interest in backlog prioritization and requirements questions. This person must be available to [illegible] [illegible]

[illegible]

Managing Extended Organizations and Data Governance

Eric Buffenoir and Isabelle Bourdon

Abstract. These last years, main IT companies have build software solutions and change management plans promoting data quality management within organizations concerned by the enhancement of their business intelligence system. These offers are closely similar data governance schemes based on a common paradigm called Master Data Management. These schemes appear generally inappropriate to the context of complex extended organizations. On the other hand, the community-based data governance schemes have shown their own efficiency to contribute to the reliability of data in digital social networks, as well as their ability to meet user expectations. After a brief analysis of the very specific constraints weighting on extended organization's data governance, and of peculiarities of monitoring and regulatory processes associated to management control and IT within these, we propose a new scheme inspired by Foucaldian analysis on governmentality: the Panopticon data governance paradigm.

Keywords: Data Quality Management, Information System Design, MDM, Community, Panopticon.

Introduction

Ten years ago, TDWI (The Data Warehousing Institute) estimated at $ 600 billion the cost of erroneous data in business sector. In fact, data quality control within an organization is a key requirement for the implementation of management control

---

Eric Buffenoir
UMR 5221 CNRS, Montpellier, France
e-mail: `buffenoir.eric@gmail.com`

Isabelle Bourdon
Montpellier Research Management, Montpellier University 2, Montpellier, France
e-mail: `isabelle.bourdon.pro@gmail.com`

P.-J. Benghozi, D. Krob, & F. Rowe (Eds.): *Digital Enterp. Des. & Manage. 2013*, AISC 205, pp. 135–145.
DOI: 10.1007/978-3-642-37317-6_12

and business intelligence. This question is all the more significant in the extended and complex organizations where differentiation between actors and organizational methods, as well as importance of external influences, strongly constrain the methods adopted to ensure consistency of standards and processes. To deal with issues of data governance, there are currently two major paradigms: Master Data Management and Community Management. The first occupies a market estimated by Gartner to $ 1.9 billion in 2012, up 21% compared to 2011 and 3.2 billion in 2015, and it is difficult to overestimate the markets covered by data quality management inherited from digital social networks.

After defining the global characters of extended organizations and clarified the specific issues of their data governance schemes, as well as the nature of the monitoring and control processes encompassed by the deployment of such governance, we address the legitimacy of existing paradigms (MDM and Community) in this context, and suggest guidelines for the development of a new data governance paradigm, better suited to the specific challenges addressed by extended organizations.

1 Data Management Issues in Extended Organizations

1.1 Extended Management, Information Systems and Data Management Issues

The notion of extended organization *(1, 2)* is characterized by the existence of multiple relationships with external partners, the delicate definition of its organizational boundaries, which become very porous, tremendous complexity of the causal dynamics in their inner evolutions, as well as nested control processes linking their various entities. Structural differentiation within extended organizations creates a peculiar need for extensive integration of their activities, which can be fulfilled by the development of transverse mechanisms and tools, crossing hierarchical chains and control, and development of multiple control channels for any process *(3, 4)*. The importance of networks in the development of cross-integration mechanisms may be preeminent over organization's hierarchical controls, due to the possible weakness of this hierarchical power on actors involved in these networks, as being exposed to strong external influences or motivated by their own interests *(5)*.

Deployment of IT puts data, their collation, processing, dissemination and quality *(6, 7)* issues at the heart of operational management control and decision-making activities *(8)*. Data governance scheme offers a framework for the definition, distribution, synchronization and exchange of reference values for Master Data *(9)*. These data are generally stored in a single place of reference, which remains in access by different applications, and: allows their creation or modification by different actors of the organization, ensures its consistent use by various operational applications, fixes a set of quality standards, facilitates the adaptation to changes of usage patterns, allows the construction of relationships between heterogeneous Master Data for decision-making processes. The implementation of a data governance scheme necessitates ***(9)***: semantic alignment between domains, clarification of concepts and identification of business glossaries, precise definition of business processes, identification of control authorities, roles and responsibilities.

We believe that the very nature of extended organizations imposes a set of technical and organizational constraints on the chosen paradigm of data governance and on the considered IS architecture, reflecting a strong incentive for decentralization of control processes over Master Data, although this decentralization may take different forms ***(10)***.

- The inherent complexity of extended organizations results in a singular complexity and a wide spectrum of Master Data, reflecting the diversity of actors, missions and organization modes for its subunits. Data governance must promote deconcentration ***(10)*** to respect the jurisdiction of actors, and multiplication / diversification of control channels on a same data set.
- Some communities within the extended organization may prefer to use their proper IS. Other inter- or trans-organizational communities may prefer integrate themselves in data governance schemes held by partner organizations and relying on their own IS tools, rather than adopting the tools and integrate the scheme coordinated by the extended organization. Hence, the pattern of data governance held by the extended organization must allow the decentralization of a significant part of control processes towards these communities and partner organizations. Considered decentralization is conceived in terms of functional decentralization or delegation ***(10)***, based on the contractual relationship between the organization and its partners, rather than in its most extreme form of devolution ***(10)***.
- Certain business processes encompassed by the data governance scheme of the extended organization inevitably involve numerous actors favouring the relations they have woven within networks over the hierarchical controls of the extended organization. The limited efficiency of these control processes does not mean the lack of normative communication concerning the quality of data among the users of these data, but rather a lack of formalization of these processes through tools, standards and processes that underpin the organization's data governance framework. This formalization may be based on the development of digital social networks and their integration in the pattern of data governance.
- Importance of external influences on the activities and resources of the extended organization constrains it to adopt standards for its data repositories that are prepared to the confrontation with the information harvested from relevant external data sources. The lack of control by the organization on IS tools used by the external data sources, imposes a systematic implementation of dictionaries between organization's Master Data and data coming from external sources.

The development of a data governance paradigm, suitable for extended organizations, raises the question of the precise nature of nested control and regulation mechanisms inherent in the use, the share and the management of data.

1.2 The Panopticon Paradigm for Data Governance

The study of monitoring and regulation mechanisms underlying management control systems and information systems has been the subject of an abundant literature. The coexistence of centralized control and empowerment of actors has been analysed in studies on control processes underlying the ERP's implementation ***(11, 12)***. These studies pointed the proximity of these mechanisms and those of the ideal control paradigm, represented by the Panopticon architecture, devised by Jeremy Bentham ***(13)*** and developed by Michel Foucault ***(14)***. In this diagram, the actor, placed in a permanent and omnipresent area of visibility, is fed continuously to act as if he was being surveiled, and is led to integrate the norms and discipline. Too rapidly identified with a regime of generalized coercion system imposed by a central authority, the panopticism is quite different from living "within a disciplinary system" ***(14)***. The panopticism is a power that does not need to manifest itself physically, to become effective, it is a "machinery that assures dissymmetry, disequilibrium, difference. Consequently, it does not matter who exercises power. Any individual, taken almost at random, can operate the machine" ***(14)***, "the Panopticon is the formula of liberal *governmentality*", this governance lies in "structuring the *field of action* of any individual by every possible ways to influence *representations*, which will play a role in the calculation of their *interests*", by acting on "*monitoring interfaces*"***(15, 16)***. The data governance paradigm within extended organizations is intended to make the considered organization a social and informational space, subject to omnipresent gaze and regulatory mechanisms. Starting in the late 80s, it was recognized how the work on Information Systems and management control ignored issues of power and conflict within organizations, and treated organizations as unified entities whose objectives are well defined and widely accepted ***(17)***. Resistance to the deployment of control processes underlying ERP within extended organizations has recently been analysed along the singular methodological approach adopted by Michel Foucault ***(18–20)***. Michel Foucault's perspective encourages to refuse the standard IS perspectives leading to analyse how the norms and the data governance scheme promoted within an organization may be legitimated and reinforced, or rather totally changed for another ones. It stands a critical method to analyse the *transformation* of control processes, which disregards schemes/institutions and the rational discourse on their own, in two ways : it privileges the study of elementary underlying disciplinary mechanisms and their articulation/discrepancy with the discursive practices ***(21)***, it also suggests to transcend the institutional standpoint and distinguish : the rationality/purpose of the institutionalizing scheme, the eventually unanticipated effects of it, the positive usage of these effects, and the formalization of a new rationality/purpose made possible by this usage and absorbing it ***(22)***. Foucault refuses to consider *institutions* as being primitive objects, fixed prior to any considerations at the same time than the collective body of individuals and their governing rules. *Institutions* are considered as focal points for the concentration of these control technologies and the production of norms, which are immediately generalized to the whole social body and circulate throughout a network woven between them, the *subject* resulting from a multiplicity of *subjugation arrangements* within them.

It is tempting to reduce Information Technologies to a global realization of the Panopticon control technology, considering the working and living environment of each individual as a space of absolute visibility for their activities *(23, 24)*, and to analyse the peculiar role played by visibility, transparency and accountability of actors in the deployment of new forms of control mechanisms permitted by IT within organizations *(25–27)* through the conceptual framework offered by Foucault. However, the isolation of the individual at the heart of the Panopticon, which makes of him "the object of information, never the subject of communication" *(14)*, is not that of the individual placed within *area of visibility* created by organization's Information System. The development of *social networks* makes him an actor of transverse communications, eventually diverting information, originally devoted to institutional control, for the purpose of strengthening the resistance of individuals to central authority *(28)*. Incidentally, the use of foucaldian analysis for data governance paradigms study relies on a shift of the standard viewpoint concerning Information Systems. These studies are focusing on *institutions* in their specific ability to fix individuals in "a place and a collective body there is no way to leave" *(29)*. To our point of view, this perspective leads naturally to translate foucaldian analysis on *institutions* for the purpose of analysing control processes promoted through Information Systems. Indeed, the very nature of information technology is to associate to objects or individuals their *digital dual* or avatar, registered in databases to proceed prescribed analysis and data matching between heterogeneous data *(30–32)*. The construction of basic business processes within the organization depends so critically on the form chosen for these digital *representations*, that the decision to develop control processes, as well as fields and methods of this control, prove to be consequences of the choice of standards and IS tools within the organization *(33)*. The digital dual is obediently and indefinitely usable for simulations coordinated by the control schemes *(34)*, as real individual is fixed to stay within foucaldian *institutions*. This *dividualization* takes then place with the consent of the real actors, driven by their interest in the use of digital tools and in the benefits of this simulation *(31)*. The participation of an actor to the control processes devoted to qualify data, relative to him and his environment, is motivated by its need to constitute himself as a *subject*, which takes shape through an act of *recognition* of its digital dual. This act of *recognition* is proceeded each time the actor is "*interpellated*" by the system (in the sense of Althusser's "interpellation" *(35)*) through *monitoring interfaces* provided by user's personal numeric environment. The precise form of these interfaces impacts deeply the efficiency of the system *(36)*. Our work will analyse the existing data governance paradigms and propose guidelines for a new paradigm directly inspired by previous considerations.

2 A New Data Governance Scheme for Extended Organizations

2.1 The Existing Data Governance Paradigm's Limits

The IT market devoted to data quality has grown through a series of relatively similar strategies and offers, entering the category of schemes called Master Data Management, that include all operations required by creation, modification or deletion of Master Data ***(37)***. The main challenge of MDM paradigm is to develop and/or strengthen processes of quality management (cleaning, de-duplication, ...) as systematically as possible ***(38)***. Thus, the analysis of business processes of the organization is a prerequisite for the implementation of this scheme ***(39)*** because the control channels, activated by a proposition to modify a Master Data, rely on the identification of data-stewards ***(40)*** with the required jurisdiction and level of responsibility to provide a level of truth to this proposal and to authorize ultimately its writing as a Master Datum (golden record). The very nature of extended organizations makes difficult the reorganization of Business Process Management (BPM) and therefore the application of the MDM scheme within them, because of

- the diversity and instability of their business processes ;
- the inefficiency of hierarchical authority over some actors of control channels promoted by the BPM, because of the prominent influence of networks and external environment ;
- the low adhesion of middle managers to issues of data quality ***(20)*** and the existence of resistance strategies from senior manager to BPM ***(20)*** ;
- the lack of control and the multiplicity of increasingly fragmented IT tools ***(41, 42)***;
- the difficulties posed by the establishment of data exchange protocols with partner organizations on a suitable collection of data ;
- the difficulties posed by the integration of data harvested from external sources.

While the MDM paradigm has nowadays established a monopolistic position on the market of data quality ***(39)***, it suffers from its inability to deal with complexity inherent to extended organizations ***(42, 43)***. To our point of view, another approach is needed in the way control processes are promoted by data governance scheme in extended organizations.

Adopting a completely opposite philosophy, another paradigm of data governance has taken a prominent place in recent years: the community paradigm that relies on self-organized online communities, oriented towards the creation and sharing of knowledge ***(44)***. The systems whose data governance model relies on this paradigm are recognized to produce data of a remarkable quality in a rather short time ***(45, 46)***. The final data (or its latest version) is the product of a social interactions process, embodied in the iterative and negotiated changes on a selected collection of data, between actors ***(45)*** within a virtual community ***(47)***. This pattern of data governance differs greatly from centralized disciplinary systems based on MDM paradigm; it relies on a democratic relativism philosophy *(48)*. Despite their efficiency, these systems remain, in our opinion, irrelevant to guarantee the conditions for deploying efficient data governance in extended organizations, because of:

- The data quality produced by the crowd in the community paradigm has been strongly criticized *(49)*. Task conflicts within the group generate both positive and negative effects on the produced content *(50, 51)*.
- The roles assigned to members within a community are self-regulated by the community, including content-oriented or administration-oriented roles *(51, 52)*. The use of self-regulated control channels and the lack of transparency and responsibility of the authors *(53)* are a major obstacle to develop data governance framework based on community paradigm in extended organizations.
- The discrepancy between priority levels assigned to a same collection of data, respectively by top-level managers of the organization and by virtual community members concerned by these data, has critical consequences on the control channels efficiency.

As a result, the Community Paradigm, cannot by itself provide a complete answer to the problem of finding a data governance scheme adapted to extended organizations.

2.2 The New Data Governance Panopticon Paradigm

MDM paradigm has been developed along the lines of preceding technical developments and existing IS architectures (ERP, BMPS, ETL, DataWarehouse). The Panopticon paradigm requires the development of new tools and architectures to articulate regulatory and disciplinary mechanisms to achieve effective data governance. This articulation is made concrete through a subtle action on representations relied on by the calculation of interests by the stakeholders, shared through their monitoring interfaces, and a control of the accountability and empowerment of the actors. This paradigm inherits main contributions from the community paradigm, but aims to compensate for its shortcomings. We propose the IS architecture of the new paradigm to be based on the existence of a specific IS element, called Panopticon IS brick, acting as a hub between existing elements of the organization's IS and personal digital environments of the individual. Thus, the fundamentals of Panopticon paradigm are the following:

- Individuals can contribute within their own customized digital environment to a set of control processes on data belonging to their field of action. The data are presented in their current state of reliability, facing the user with the interpellation of the system to recognize its digital dual world and then constitute himself as a subject by using its power to tell their truth on these data. Unlike MDM solutions working downstream of IS elements, like a Extract-Transform-Load (ETL) system acts towards a Data Warehouse, Panopticon IS brick maintains its reference databases through real-time processes.

- Complete transparency and traceability are ensured on the set of required interventions made from individual actions or external sources (proposition to change the value of a given reference datum, reasoned opinion emitted to conclude within a given control channel, arbitration control between divergent control channels). Each actor involved in a control channel is then placed in an area of visibility for an invisible community of actors, concerned by the same data, in order to promote self-discipline and integration of norms. However, anonymity can be ensured on free contributions devoted to the warning about erroneous data and critical/ranking processes, in order to promote emancipation of the individual with respect to the issue of managing data.
- This approach is user-centric, in the sense that the collection of reference data, covered by the data governance scheme, is chosen according to the set of data used by the set of digital services offered to the users. User-interface is constantly adapted to the currently used services in order to optimally leverage personal interest of the users to get them to participate to control processes. This interest relies on its need to access services based on up-to-date and personalized data, to cooperate with other members of his networks, to develop competitive strategies to access shared resources, or to exercise his responsibilities.
- Numerous control channels exist for any given datum, a control channel is indeed associated to any community concerned by the different usages of this datum. Each control channel is formalized by the allocation of structuring roles and prioritized rights about this datum to any individuals within this community: rights to read, rights to freely warn for an erroneous data, rights and responsibility to propose a modification of a datum, rights and responsibility to evaluate/control the propositions to change a datum made by other individuals, right and responsibility to arbitrate between divergent controls. The set of control channels formalized by the system encompass the whole set of ties, controls or transactions, inherited from networks and coalitions existing within the organization, as well as conflictual and competitive relationships, although these relations are generically transverse to hierarchical relationships of the organization.
- Unlike in MDM scheme where the control channels are initial parameters for the system, the Panopticon paradigm allows the communities to self-organize the control channels. This bias is imposed by the objective fixed by the system to take into account the complex dynamics of these networks. Modifications made by an individual, on the hierarchical data belonging to its field of action, contribute to change this field of action, as well as the area of visibility within which he is located, but also to modify or constrain those of the other individuals. In order to conciliate the multiplication of self-organized control channels and the efficiency of the whole control process, we have to impose basic requirements: unlike in community-based data governance schemes a unique control channel associated to hierarchical channel inherits the arbitration power on the final decision and responsibility to change the golden record, the whole set of control channels concerned by the same collection of data are ranking/censoring/granting each other according to the rights they

have to act on hierarchical data corresponding to the details of the other control channels.

- While the MDM paradigm is not well adapted to the integration of external data sources, they should be extensively used by Panopticon scheme. They must be considered as well as the control channels emerged from communities to anticipate improvements and remedy to the control processes, which do not meet the appropriate data quality threshold. Control channels and data sources are subject to a ranking process by comparison with the results of other channels.
- The answer given by the MDM/ERP paradigm to the issue of fostering data exchange protocols between the organization and its partners is to impose a single integrative framework for business processes. By contrast, MDM paradigm neglects the existence of internal boundaries emerging within organizations from resistance strategies deployed by some of its sub-units. To deal with these two types of boundary problems, the strategy adopted by Panopticon paradigm should be to promote a "functional decentralization" of a significant part of the control processes through the development of a distributed IS architecture based on numerous instances of the Panopticon IS brick. This strategy promotes the dissemination of norms underlying the reference databases of the Panopticon IS brick, at the cost of losing visibility on a part of control processes carried out within the subunits.

3 Conclusion

After having clarified the constraints on data governance schemes within extended organizations, it became apparent that the current paradigms underlying the Master Data Management solutions, or adopted by digital networks communities, do not meet them. An analysis of the regulatory and disciplinary controls within these extended organizations has led us to propose a new paradigm to meet the constraints weighting on the deployment of such a scheme, it requires technological developments that should be the object of a specific research.

References

1. O'Neill, H., Sackett, P.: The Extended Manufacturing Enterprise Paradigm. Management Decision 32, 42–49 (1994)
2. Browne, J., Sackett, P.J., Wortmann, J.C.: Future manufacturing systems—Towards the extended enterprise. Computers in Industry 25, 235–254 (1995)
3. Lawrence, P.R., Lorsch, J.W.: Differentiation and Integration in Complex Organizations. Administrative Science Quarterly 12, 1 (1967)
4. Galbraith, J.R.: Designing Complex Organizations. Addison-Wesley Longman Publishing Co., Inc., Boston (1973)
5. Lavigne, A.: La communication institutionnelle vue par l'entreprise de six grands – Systèmes relationnels: proposition d'une typologie. Communication et organisation (2002)

6. Wang, R.Y., Strong, D.M.: Beyond accuracy: what data quality means to data consumers. J. Manage. Inf. Syst. 12, 5–33 (1996)
7. Batini, C., Cappiello, C., Francalanci, C., et al.: Methodologies for data quality assessment and improvement. ACM Comput. Surv. 41, 16:1–16:52 (2009)
8. Drucker, P.F.: Management Challenges for the 21st Century. Harper Business (2001)
9. Régnier-Pécastaing, F., Gabassi, M., Finet, J.: MDM: Enjeux et méthodes de la gestion des données, Dunod (2008)
10. Lemieux, V.: Decentralisation, Politiques Publiques et Relations de Pouvoir. PUM (2001)
11. Sia, S.K., Tang, M., Soh, C., et al.: Enterprise resource planning (ERP) systems as a technology of power: empowerment or panoptic control? SIGMIS Database 33, 23–37 (2002)
12. Elmes, M.B., Strong, D.M., Volkoff, O.: Panoptic empowerment and reflective conformity in enterprise systems-enabled organizations. Inf. Organ. 15, 1–37 (2005)
13. Bentham, J.: Panopticon or The Inspection-House. Miran Bozovic, London (1995)
14. Foucault, M.: Surveiller et punir, Gallimard (1975)
15. Foucault, M.: Naissance de la Biopolitique. Gallimard-Seuil (1979)
16. Foucault, M.: Le sujet et le pouvoir, Dits et Écrits, pp. 208–226. Gallimard (1982)
17. Robson, K., Cooper, D.J.: Power and management control. In: Critical Perspectives in Management Control, Wai Fong Chua, Tony Lowe and Tony Puxty, pp. 79–114. Macmillan (1989)
18. Doolin, B.: Information technology as disciplinary technology: being critical in interpretive research on information systems 13, 301–311 (1998)
19. Clegg, S.R., Courpasson, D., Phillips, N.: Power and Organizations (2006)
20. Beckett, C.: Information Systems, Power and Resistance: A Genealogical Inquiry Into Failure, The Information System Stream, Naples (2011)
21. Ezzamel, M., Delbridge, R.: The Strength of Difference: Contemporary Conceptions of Control. Organization 12, 603–618 (2005)
22. Foucault, M.: Dits et Écrits 1976-1988, Gallimard, Paris, France (1984)
23. Robins, K., Webster, F.: Cybernetic Capitalism: Information, Technology, Everyday Life. In: The Political Economy of Information, pp. 47–75. Vincent Mosco & Janet Wasko, Univ of Wisconsin Press (1988)
24. Zuboff, S.: In the age of the smart machine: the future of work and power. Heinemann Professional (1988)
25. Miller, P., O'Leary, T.: Accounting and the construction of the governable person. Accounting, Organizations and Society 12, 235–265 (1987)
26. Willcocks, L.: Foucault, power/knowledge and information systems: reconstructing the present. Wiley, Chichester (2004)
27. Klecun, E.: Information and Communication Technologies (ICT) as Control Mechanisms. Presented at the CMS7 2011 - 7th International Critical Management Studies Conference, Faculty of Economics. University of Naples Federico II - Naples (2011)
28. Witheford, N.: Cycles and Circuits of Struggle in High Technology Capitalism. In: Davis, J., Hirschl, T.A. (eds.) Cutting Edge: Technology, Information Capitalism and Social Revolution, pp. 195–242 (1997)
29. Legrand, S.: Les normes chez Foucault. PUF, Paris, France (2007)
30. Clarke, R.: The digital persona and its application to data surveillance. The Information Society 10, 77–92 (1994)
31. Poster, M.: Databases as Discourse, or, Electronic Interpellations. In: Lyon, D., Zureik, E. (eds.) Computers, Surveillance, and Privacy, p. 175. University of Minnesota Press (1996)

32. Latour, B.: Drawing things together. In: Dodge, M., Kitchin, R., Perkins, C. (eds.) The Map Reader: Theories of Mapping Practice and Cartographic Representation (2011)
33. Lyon, D.: Electronic Eye: The Rise of Surveillance Society. University of Minnesota Press (1994)
34. Lyon, D.: Surveillance society: Monitoring Everyday Life. Open University Press (2001)
35. Althusser, L.: Idéologie et appareils idéologiques d'État. La pensée 151, 3–38 (1970)
36. Simon, B.: The return of panopticism: Supervision, subjection and the new surveillance. Surveillance & Society 3 (2002)
37. Smith, H.A., McKeen, J.D.: Developments in Practice XXX: Master Data Management: Salvation Or Snake Oil? Communications of AIS, 63–72 (2008)
38. Otto, B., Hüner, K.M., Österle, H.: Toward a functional reference model for master data quality management. Information Systems and e-Business Management 10, 395–425 (2011)
39. Friedman, T., Bitterer, A.: Magic Quadrant for Data Quality Tools. Gartner Inc. (2012)
40. Loshin, D.: Master Data Management. Morgan Kaufmann (2009)
41. Dahlberg, T., Heikkilä, J., Heikkilä, M.: Framework and Research Agenda for Master Data Management in Distributed Environments. In: Leino, T. (ed.) Proceedings of IRIS 2011. TUCS Lecture Notes, vol. 15, pp. 82–90 (October 2011)
42. Silvola, R., Jaaskelainen, O., Kropsu-Vehkapera, H., et al.: Managing one master data – challenges and preconditions. Industrial Management & Data Systems 111, 146–162 (2011)
43. Otto, B., Reichert, A.: Organizing master data management: findings from an expert survey. In: Proceedings of the 2010 ACM Symposium on Applied Computing, pp. 106–110. ACM, New York (2010)
44. Lee, G.K., Cole, R.E.: From a Firm-Based to a Community-Based Model of Knowledge Creation: The Case of the Linux Kernel Development. Organization Science 14, 633–649 (2003)
45. Korfiatis, N.T., Poulos, M., Bokos, G.: Evaluating authoritative sources using social networks: an insight from Wikipedia. Online Information Review 30, 252–262 (2006)
46. Liu, J., Ram, S.: Who Does What: Collaboration Patterns in the Wikipedia and Their Impact on Data Quality. SSRN eLibrary (2009)
47. Balasubramanian, S., Mahajan, V.: The Economic Leverage of the Virtual Community. Int. J. Electron. Commerce 5, 103–138 (2001)
48. Grassineau, B.: Wikipédia et le relativisme démocratique. Observatoire des Mondes Numériques en Sciences Humaines (2007)
49. Lipczynska, S.: Power to the people: the case for Wikipedia. Reference Reviews 19, 6–7 (2005)
50. Viégas, F.B., Wattenberg, M., McKeon, M.M.: The Hidden Order of Wikipedia. In: Schuler, D. (ed.) HCII 2007. LNCS, vol. 4564, pp. 445–454. Springer, Heidelberg (2007)
51. Arazy, O., Nov, O., Patterson, R., et al.: Information Quality in Wikipedia: The Effects of Group Composition and Task Conflict. Journal of Management Information Systems 27, 71–98 (2011)
52. Ren, Y., Kraut, R., Kiesler, S.: Applying Common Identity and Bond Theory to Design of Online Communities. Organization Studies 28, 377–408 (2007)
53. Santana, A., Wood, D.J.: Transparency and social responsibility issues for Wikipedia. Ethics and Information Technology 11, 133–144 (2009)

From Organization Design to Meta Organization Design

Rolande Marciniak

Abstract. Organization design can be a fruitful inspirational source when doing research on contemporary organizations. The star model's first version was adapted so it matches the evolution of firm strategy. These adaptations far from signaling the star model's weakness show the real strength of a systemic framework where some of the firm's essential components and design evolutivity are built-in. Since the 1980s the increase in close collaboration between formally independent firms and legally autonomous actors poses challenges for our thinking about organizational design. If Meta-organizations involve multiple firms as well as communities of non-contractually linked individuals, an emphasis on intrafirm design may be incomplete. Because firms have partially moved from stand-alone organizations to meta-organizations, we propose to include meta-organization design enabling firms to collaborate.

1 Introduction

The world of organizations has evolved since the foundational theories of organizational design were first postulated. From the analysis of the seminal work by J. R. Galbraith (1973) organization design can be a fruitful inspirational source when doing research on contemporary organizations. The star model's first version was adapted so it matches the evolution of firm strategy. Over the course of the twentieth century, because of their environment's complexity and hostility, large firms adopted a complex organization (Galbraith, 2010). These adaptations far from signalling the star model's weakness show the real strength of a systemic framework where some of the firm's essential components and design evolutivity are built-in.

Rolande Marciniak
Université Paris Ouest
Nanterre La Défense, Nanterre, France
e-mail: rolande.Marciniak@wanadoo.fr

P.-J. Benghozi, D. Krob, & F. Rowe (Eds.): *Digital Enterp. Des. & Manage. 2013*, AISC 205, pp. 147–159.
DOI: 10.1007/978-3-642-37317-6_13

Since the 1980s the increase in close collaboration between formally independent firms and legally autonomous actors poses chall enges for our thinking about organizational design. If new forms of organization increasingly involve multiple firms as well as communities of non-contractually linked individuals, an emphasis on intrafirm design may be incomplete (Gulati and al., 2012). Because firms have partially moved from stand-alone organizations to meta-organizations, we propose to include the new organization design that enables firms to collaborate.

The chapter is organized as follows. First, organization design and more specifically the star model (2.1.) and its contemporary applications (2.2.) are presented. Second, the notion of meta-organization and its dimensions are analysed (3.1) to introduce some principles, given the early stages of these new forms of organizing, concerning meta-organizations design (3.2.).

2 Organizational Design

OD makes up one of the different brands in organization theory. OD theory appeared between 1973 & 1978 and its most famous figures comprise J. R. Galbraith, R. E. Miles, and C. C. Snow. This particular group has outlined the components of an organisation viewed as an open social system and shown that an organization may prove efficient only when its different sub components interact with each other in a congruent fashion. This part focuses more especially on J. R. Galbraith's star model starting with a presentation followed by the model's recent applications.

2.1 The Star Model

Following up on his initial work akin to the research by the firm's behavioral school, J.R. Galbraith (1973) worked out one model (the star model) that allows business leaders to conceive organization patterns which can positively affect behaviour. This presentation of the star model draws on Galbraith's most recent and generic book (2002). The firm's main components are examined.

The strategy targets the firm's objectives, values and assignments. It shows which path to follow up as it indicates what is on offer (products/ services) and which markets to be supplied. Thus the strategy provides the basis on which to rest the structural choices which inevitably include some way of compromise.

The structure targets the differentiation among the different units. Four sets of criteria are at play when setting up the structure, i.e., specialization, subordination range, power distribution (centralisation vs. decentralization), departmentalization. The units may be based upon on the function, product or geographic area etc. Whatever the structure — functional, divisional, matrix, or hybrid — all of them present their advantages and drawbacks.

The processes consist of the information and decision making fluxes running vertically and horizontally across the organization. Vertical fluxes allocate resources and usually include budgeting and forward planning. Horizontal fluxes aim to achieve units' integration. Five process categories may be put into force whenever the structure wants some flexibility: informal groups, artefacts, ie coordination through ICT (Information and communication technologies) and IS (information systems), formal groups, integrator managers positions and finally, the matrix structure. Besides, the process categories are not mutually exclusive.

The motivation system aims to align personnel goals with those of the firm. They provide the required motivation to move in the chosen direction and meet the strategy's objectives. It covers financial aspects (salaries, bonuses and allowances, stocks etc.) as well as other perquisites such as company car and parking facilities etc.

Personnel policy comprises hiring, internal mobility, training, and promotes the strengthening of the skills required for setting up the strategy.

As Galbraith put it "Organizational design is the search for coherence between strategy (domain, objectives and goals), organizing mode (decomposition into subtasks, coordination for the completion of whole tasks) integrating individuals (selection and training of people), and designing a reward system." (Galbraith, 1977, p. 5) The conception approach first specifies the criteria originating in the strategy and then moves on to the choice of structural criteria. Once the structure that best fits the strategic goal has been selected, key processes are set up with a view to providing more flexibility to the organization's functioning. The approach is then carried on by selecting the key managers and defining their role and responsibilities. The various IS are then examined along with the incentive system and the performance evaluators. Finally, training, career management and development are taken care of. Although the approach is sequential, it is often necessary to go back and forth. Moreover, it is suggested that whenever the strategy's outline is ill defined the processes should be tackled first because they're more flexible compared to the more rigid structure and a total overhaul is never a good idea. Organisation design is an on-going managerial activity as on the one hand the current organisation indeed needs upraising to make sure it is consistent with the environment and upgraded when necessary. On the other hand, the organisation of the future has to be groomed if tomorrow's strategy is to succeed.

2.2 *Current Relevance of the Star Model*

The star model's first version goes back to 1973. It was adapted so it matches the evolution of firms. Thus five versions deal with: (1) the innovating firms (Galbraith, 1982), (2) the firms that are subject to a hyper dynamic global environment (Mohrman, Galbraith and Lawler, 1998), (3) the global corporation (Galbraith, 2000), (4) the client oriented firms (Galbraith, 2005) and, (5) the multidimensional reconfigurable organization (Galbraith, 2010). These adaptations far from signalling the model's weakness show the real strength of a systemic framework where some of the firm's essential components and design evolutivity are built-in with specific elements for each version.

First, J. R. Galbraith's recent findings and then two recent OD-inspired studies are presented.

Because of their environment's complexity and hostility, large firms adopted a **multidimensional reconfigurable organization** at the turn of the 2000s. The more dynamic the environment, the more frequently the reconfiguration of internal and external relationships should occur. In response, leading firms in complex, dynamic environments are experimenting with reconfigurable organization structures (Galbraith, 2010). The main characteristic of this organizational complexity is a multi-dimensional matrix structure on the one hand, and the setting up of a two-faced organization on the other, the first being more stable while the second is reconfigurable and in constant evolution. The organisation's variable parts comprise the teams that are constantly forming and re-forming on the one hand and on the other, the decision committees which allocate the resources and decide on the priorities. The reconfigurable part is made and unmade to better seize on and work out the planned or emerging opportunities. The decision processes are made of a stable core of participants on the one hand and on the other of a flow of participants renewed according to which opportunities need addressing. The integration of the reconfigurable multi-dimensional structure is achieved through various devices built in the star model's different components. The strategy's own pace is imposed on other components. As it indeed sits on the organisation's border, the strategy reflects the environment's complexity, dynamism and hostility perceived by the business leaders. The main obstacle lies in the components' ability to adapt to a complex strategy: "… many companies today are trying to pursue strategies that far exceed the capabilities of their organizations… …companies are pursuing third generation strategies using second generation organizations that are staffed with first generation human resources… We need to invest in and develop the capabilities of our people and organizations before we can master today's complex global economy." (Galbraith, 2010, p. 124) All coordination-integration forms are widely developed and supported by ICT. Staff policy involves a highly selective recruitment procedure ("hire hard, manage easy"), brisk internal mobility, training programs — which include meticulous drilling in how to run the complex and reconfigurable organisation.

The star model has also been used over the last ten years for studying its impact on **call centers' productivity** (Rowe, Marciniak and Clergeau, 2011) and the evolution of **knowledge management during a merger** (Ben Chouikha, Marciniak, 2013). We move away from the star model on two important points. Unlike Galbraith's, our objective was not geared towards the setting up of organization units but the gathering within an integrating model of the elements often scattered away and originating from research prior to ours, it aimed to test the star model's framework in a contextualized field and to study the coherence between the model's different components along with its impact on performance. This abuse of objectives appears fruitful to us in so far as it provides an integrating framework that avoids dispersal and tests the model in different sectors and contexts. The second difference concerns the component called processes by Galbraith. It actually appeared essential to us that the distinction should be drawn between IT and communication processes. Although Galbraith looks on them as one single component, they make up two different constructs. As integrating units and coordinating tasks are increasingly supported by tools, it is becoming important to distinguish their relative impacts since no organization can currently ignore these tools and therefore all organizations have, to some extent, been turned into a digital organization. This distinction between process and technology has been clearly formulated in BPR studies (Davenport, 1998; Clark and Stoddard, 1996). In order to justify our deviation, it must also be noted how much Galbraith insists on the importance of tooled processes (ERP, Workflow, Project management systems etc.) and consequently ascertains the IS's prominent role.

Several decades after the star model came up, it still proves pertinent to either work out an organization's structure or study the quality and the specificity of the arrangement of the model's components in varied contexts. We believe that distinguishing between the information technology and other processes has become a must when the objective aims to conceive or re-conceive an organization because the tools likely to make it more efficient cannot be ignored, and/ or to study the tools' impact and how they link up with the model's other components.

3 Meta-Organization Design

Beginning in the late 1990s and continuing to the present, firms began to move toward a new business model housed in a new organizational form, a form that incorporated both independent firms and their networks as building blocks. Within some industries, firms are currently exploring the community model for the purpose of assuring the full utilization of continuously developing knowledge (Miles et al., 2009; Fjeldstag et al., 2012). Other authors (Gulati, et al. 2012) point to an important phenomenon, the emergence of *meta-organizations*, having two manifestations. Since the 1980s, firms have entered into collaborative relationships; such partnerships typically span geographies, industries, and value chains. The rise of strategic outsourcing more broadly and business process outsourcing more particularly is another indicator of the strength of this phenomenon. A second manifestation is that the Internet and related technologies have become tools of both knowledge production and dissemination; this hastens the recognition that actors outside the traditional boundaries of the firm possess unique knowledge that may be applicable within the firm. The growing capacity for geographic work dispersion facilitated by communication and information technologies is an important determinant of meta-organizations. Catalyzed by falling communication costs, many organizations have developed sophisticated practices that enable the division of labor and the reintegration of efforts across geographies in ways that were inconceivable a few decades ago.

We first present the meta-organization and its dimensions (3.1.) then we work out a first draft for meta-organization design (3.2.).

3.1 *Meta-Organization (M-O)*

The term meta-organization is used by Ahrne and Brunsson and Gulati and al. but we also associate to our work the research on collaborative communities, network firms, business ecosystems, and interorganizational relationships.

3.1.1 Definition and Features of Meta-Organization

For Ahrne and Brunsson (2005) meta-organizations are organizations whose members are other organizations (firms, states or associations). They include well-known examples such as the United Nations or the Fédération Internationale de Football (FIFA), as well as others less well-known.

Gulati and al. (2012, p. 573) define meta-organization as "… *an organization whose agents are themselves legally autonomous and not linked through employment relationships. An agent in this definition could itself be an organization (within which there may well be employment relationships), but which can be treated as a unitary actor for purposes of analysis. Thus, meta-organizations comprise networks of firms or individuals not bound by authority based on employment relationships, but characterized by a system-level goal.*"

This second definition, including individuals as possible participants of meta-organizations, seems more interesting for analyzing contemporary on line M-O and social networks comprising the multitude of human beings.

Meta-organizations (M-O) resemble biological super-organisms comprising a multitude of individual organisms that coexist, collaborate, and coevolve via a complex set of symbiotic relationships which together form a larger organism. But, even if the system-level goal of an M-O can be emergent, as it is the case with purely self-organized systems, more often it corresponds to the goals of the architects of the M-O. For instance, there is no doubt that leading companies — such as P&G, ARM Holdings Plc, Dassault Systèmes, Apple or Google —have gained success by powerfully shaping (although not fully determining) the formation of the M-Os surrounding them (Williamson and De Meyer, 2012).

M-Os represent a particular kind of organizing relations among legally autonomous entities where central actors resort to substitutes for formal hierarchical authority when fashioning the M-O's design. They rest on expertise, reputation, status, access to resources.

There are other additional features besides the absence of formal authority, though not unique to M-Os. The incentive system other than financial plays an important part where most essential are the incentives stemming from intrinsic motivation, needs, and reputations. For instance, these self-motivated, self-selected and self-governed communities (Boudreau and al., 2011) have dramatic results in problem solving (Lakhani and Von Hippel, 2003). In the M-O, resorting to ICT technologies together with the partitioning of tasks that allows independence are substitutes for colocalization (Srikhanth and Puranam, 2010).

3.1.2 Dimensions of Meta-Organizations

Different alternatives to M-Os are generated and patterns within this variation may be understood by examining important dimensions of M-Os.

Permeability of boundaries, degree of stratification (Gulati and al., 2012)
Deliberation about the extent and limits of purposive M-O shapes the attraction and retention of its members. **Boundary arrangements** include: criteria for membership, duration and exclusivity of membership, how membership decisions are made, and how members' contributions are controlled. Closed boundaries are reminiscent of strategic alliances with an explicit and tailored definition of tasks, and are associated with fewer members. Open membership makes the timing of members' entry and exit difficult to control. They can result in unsolicited and unwanted contributions as well as in contestations of collective goals and agreements. Decisions about the boundaries and the openness of membership alter the behavioral dynamics within M-Os, as well as the range of feasible governance arrangements. The **degree of stratification** concerns the differentiation in the roles of membership. Stratification helps reduce complexity by subdividing the M-O into smaller groups and exploiting the innate hierarchy of tasks (Simon, 1962). Stratification can also serve as a motivational device. The material and symbolic benefits associated with higher status or role-based authority can create incentives for M-O members. A low degree of stratification is likely to support the emergence of a community of equals who are more likely to adopt peer-based approaches to coordination. Minimizing stratification within M-O also permits to avoid, for example, some debates about the criteria and processes for determining status, and encourage widespread participation. Where broad participation based on involvement and identification with M-O is critical, lower degrees of stratification may be preferable. Decisions about stratification within M-Os have significant impacts on both motivation and coordination.

Purposes, actors, infrastructures-protocols-processes, commons (Fjeldstad et al., 2012)
An actor-oriented scheme comprises four dimensions of M-Os: purposes, actors, protocols-processes-infrastructures, and commons. The specific **purposes** are "la raison d'être" of the M-O. For instance, established in 2006 by IBM and seven other founding firms, Blade.org is a collaborative community of more than 200 firms whose purposes are the development, manufacturing, marketing, and distribution of solutions based on the blade server technology invented by IBM. Rather than attempting to exploit its blade technology through its own business units, IBM chose to form a meta-organization for accelerating the adoption of blade server solutions. **Actors** –organizations, associations, individuals–, members of the M-O, have the capabilities and values to accomplish the purposes. For example, Blade.org membership comprises 70 complementary firms representing the different capabilities required to develop solutions and 180 firms that are their customers. **Protocols processes and infrastructures** enable multi-actor

collaboration. Protocols are codes of conduct used by actors in their collaboration activities. A category of protocols deals with the division of labor. Other protocols deal with inter-actor coordination. For instance Blade.org has 9 technical committees, and solutions are developed through 4 forms. Infrastructures, or technological platforms, are systems that connect actors. **Commons** refers to resources (knowledge, common awareness, code source, data…) that are collectively owned and available to the actors. For example commons comprise standards, solutions, shared situational awareness and tracking of property rights. Taken together, these elements create and function within contexts consisting of various combinations of transparency, shared values, norms of reciprocity, trust, and altruism.

Control, interdependence (Koenig, 2012)
The centralised or non-centralised **control** over the ecosystems' essential ressources allows us to distinguish between M-Os that are controlled by a leader firm from those that are not. Control in the firms' networks translates into a continuum varying from an asymetrical network to a symmetrical one (Gandori and Soda, 1995). The **type of interdependence** linking up M-O members influences the size of the network and its topology. When one moves from pool interdependence to reciprocal interdependence and when demands on communication coordination increase, the number of actors likely to interact directly decreases. The actors' interdependence conditions the M-O's development mode: pool interdependence favors a multi dimensional quantitative development whereas reciprocal interdependence facilitates the qualitative development based on deepening relations.

Platform, business model, leadership (Edouard and Gratacap, 2011)
Business ecosystems (BE) are structured around three elements: the technological platform, the business model and leadership. The **technological platform** is a set of technical solutions and services available to each M-O member. The platform makes up the BE'structuring architecture. It is the device with which the pivot firm organizes value creation by offering manyfold opportunities and collecting many types of contributions. It also distributes value among the different members (Iansiti and Levien, 2004, Evans and Zchmalensee, 2007). Platform strategies multiply on dual faced markets – video games platforms put players and game developers into contact — or multi faced whereby the Amazon Webservice platform brings together publishers, bookstore keepers, reviewers, readers and application developers. For its own part, Microsoft develops its platform according to the number of communities it wishes to target: PC for Windows, Cloud computing for Azure and video games for Xbox. Different platform categories can be distinguished : those that allows members to exchange —eBay, Amazon , those which bring into contact an audience likely to be valorized by it — Google, the platforms which produce products or services generating indirect network externalities – Blade.org, games platforms (Iskia, 2011). The **business model** incorporates partner networks as an element structuring M-Os. « Having a business model for your firm is not enough. Executives must become ultra-sophisticated at developing business models for their respective communities. » (Moore 1996, p. 57). A business model only performs if it creates value and if it manages to capture some of that

value. Its objective is to establish consistency between creating and distributing value among partners — for each dollar earned by Microsoft $8.70 are earned by its partners. The business model decribes the parts, the relations and the partners' flux of information, knowledge, skills, revenue and products. **Leadership** is a competitive advantage and is gained over time through the firm's aptitude to combine key competences and to share common values.

3.2 Meta-Organization Design

The term of architecture constitutes the synthesis of form in response to function. Structure should be consistent with the purpose activity system as form must follow function. An M-O is a finalized system that is effective if it has been designed in a coherent manner. Meta-organization designs are emerging in which rich sets of resources are made available to a large set of actors who self-organize on unlimited sets of projects. Reliance on self-organization and local decision making in the development of products & services requires mechanisms that allow actors to become aware of problems and opportunities and identify and form relationships with suitable co-actors. The collaborating parties must be able to manage their common resources and goals and overcome the agency problem of free riding. The lateral nature of decisions about which projects to pursue, which resources to share, and how returns will be divided is a preeminent characteristic of the architecture of meta-organization forms.

We propose six components of the M-O design: purposes, membership & governance, actors, structures & processes, technological platform, and commons.

3.2.1 Purposes

Here we mean the overall objective of the meta-organization's assignment, the reason why it exists. This assignment was initially formulated by its founders, yet it may evolve and its evolution is taken into account in the governance principles. Purposes guide the design of membership and governance principles.

3.2.2 Membership and Governance Principles

A majority of M-O can be classified using the predominant means of participation —closed vs open— and the predominant governance structure —hierarchical vs flat— (Pisano and Verganti, 2008). The options concerning membership are: who decides on the selection criteria, who select the actors, the duration for participation. The option choices depend on the assignment and impact the meta-organization's structure. The governance principles provide regulation rules for the M-O's evolution and running. They aim to guide the actors' behavior to make sure their specific goals are congruent with those of the meta-organization. The governance principles cover the following topics. How are the M-O's investment and running costs financed? Where in the M-O is value created and how is this value captured and shared? Members' code of conduct (rights and duties) Meta-rules concerning the evaluation and decision-making processes, the management of commons. M-O's evolution system —of which the scalability or the adaptation to the rise of the meta-organization.

3.2.3 Actors

A status-based actor's taxonomy is necessary so as to conceive relevant incentive systems and avoid inconsistencies. For instance, Blade.org has five types of actors: (1) founding firms, (2) firms developing and distributing hardware, software, or services for the blade platform, (3) firms providing consulting or distribution support for blade-based solutions or products, (4) firms using blade platform solutions, and (5) firms being customers or end users. Fashioning the M-O's design implies using substitutes for formal hierarchical authority such as expertise, reputation, status, access to resources and the stratification of members helps the M-O design.

3.2.4 Structure and Processes

Organizing involves dividing and integrating resources in order to control and coordinate activities (Mintzberg, 1983).The meta-organization seems to be functioning with a structure lighter than that of other organizations. Membership & stratification of actors are guides for dividing activities inside an M-O. Governance principles help to design M-O processes and most of them are incorporated in the technological platform. M-O organization evolves over time by adding news ways for members to collaborate or inputs into the decisions process. Structures can be patterned on what the architects consider the best practices and on what members need to meet certain task requirements. Three basic features of the M-O structures seem to be durable: a light team of architects working closely together, few levels of management under the team and, individuals organized into project teams work within designated areas.

3.2.5 Technological Platform

The platform coordinates actors and favors collective value creation by sharing resources. Coopetition management is supported by the platform's strategy which involves architectural choices and setting up the platform. Thus technological bricks can be integrated into a new offer so they allow the M-O to explore several trajectories and expand its network. The M-O's rules of the games are either partially or totally incorporated into the platform. The ecosystem members' participation are likely to prompt the platform to change as they open up a certain number of strategic options. The platforms may indeed develop the M-O's market spaces through two levers : (1) depth by creating new product/ service functionalities that will saturate the existing clients' needs and (2) the spread by seeking new value source through adding new faces or recruiting new economic agents' communities on the platform.

3.2.6 Commons

All manner of resources both tangible and intangible can be shared by the M-Os' members. Those ressources' management rules lie within the M-O's governance principles. Besides, many common resources are managed through the structures and processes and integrated into the technological platform.

4 Conclusion

Born in the early 1970s, the organization design stream proved robust and flexible enough to conceive the organizations in spite of their economic and technical environment's important evolution. To this day the star design model's principles remain relevant to these organizations' conception and analysis.

However from the 1980s onwards different research fields like network firms, strategic alliances and business ecosystems etc. have underlined the importance of developing inter-firm relations. But scholarly work on the various forms of multi assemblages has been disconnected. Each era's theories, in part reflects the managerial preoccupations of the times and coevolves with them. Because to day, a lot of organizations are involved in close collaboration while their managers wonder about the potential strengths and weaknesses of this strategy, an integrative objective of conceptualizing all these types of collectives as meta-organizations constitutes at the present time a relevant research programme. This programme has to treat a cluster of legally autonomous entities as an organization and must acknowledge that such a M-O, like any organization, embodies key structural elements that can be designed in a coherent manner.

References

Ahrne, G., Brunsson, N.: Organizations and meta-organizations. Scandinavian Journal of Management 21, 429–449 (2005)

Ben Chouikha, M., Marciniak, R.: Le management des connaissances: le cas des équipes projet Nokia Siemens Networks en Tunisie. Système d'Information et Management (2013, forthcoming)

Boudreau, K.J., Lacetera, N., Lakhani, K.R.: Incentives and problem uncertainty in innovation contests: an empirical analysis. Management Science 57(5), 843–863 (2011)

Clark, T., Stoddard, D.: Interorganizational business process redesign: merging technological and process innovation. Journal of Management Information Systems 13(2), 9–28 (1996)

Davenport, T.: Putting the enterprise into the enterprise system. Harvard Business Review 76(4), 121–131 (1998)

Edouard, S., Gratacap, A.: Proposition d'un modèle d'intelligence collective pour les écosystèmes d'affaires. Management & Avenir 46(6), 177–199 (2011)

Fjeldstad, O.D., Snow, C.C., Miles, R.E., Lettl, C.: The Architecture of Collaboration. Strategic Management Journal 33, 734–750 (2012)

Galbraith, J.R.: Designing complex organizations. Addison-Wesley Publishing Company (1973)

Galbraith, J.R.: Organization Design. Addison-Wesley Publishing Company (1977)

Galbraith, J.R.: Designing the innovating organization. Organizational Dynamics 10(3), 5–25 (1982)

Galbraith, J.R.: Designing The Global Corporation. Jossey Bass (2000)

Galbraith, J.R.: Designing Organizations an Executive Guide to Strategy Structure and Process. Jossey Bass (2002)

Galbraith, J.R.: Designing the Customer-Centric Organization a Guide to Strategy, Structure, and Process. Jossey Bass (2005)

Galbraith, J.R.: The Multi-Dimensional and Reconfigurable Organization. Organizational Dynamics 39(2), 115–125 (2010)

Gandori, A., Soda, G.: Inter-firm networks: antecedents, mechanisms and forms. Organization Studies 16(2), 183–214 (1995)

Gulati, R., Singh, H.: The Architecture of Cooperation: Managing Coordination Costs and Appropriation Concerns in Strategic Alliances. Administrative Science Quarterly 43, 781–814 (1998)

Gulati, R., Puranam, P., Tushman, M.: Meta-Organization Design: Rethinking Design in Interorganizational and Community Contexts. Strategic Management Journal 33, 571–586 (2012)

Iansiti, M., Levien, R.: Strategy as ecology. Harvard Business Review, 69–78 (March 2004)

Iskia, T.: Ecosystèmes d'affaires, stratégies de plateforme et innovation ouverte: vers une approche intégrée de la dynamique d'innovation. Management & Avenir 46(6), 157–176 (2011)

Koenig, G.: Le concept d'écosystèmes d'affaires revisité. M@n@gement 15(2), 208–224 (2012)

Lakhani, K.R., Von Hippel, E.: How open source software works: "free" user-to-user assistance. Research Policy 32(6), 923–943 (2003)

Miles, R.E., Snow, C.C., Fjeldstad, O.D., Miles, G., Lettl, C.: Designing organizations to meet 21st-century opportunities and challenges. Organizational Dynamics 39(2), 93–103 (2010)

Mintzberg, H.: Structure un fives. Prentice Hall, Englewood Cliffs (1983)

Mohrman, S.A., Galbraith, J.R., Lawler, E.E.: Tomorrow's Organization Crafting Winning Capabilities in a Dynamic World. Jossey-Bass Publishers, San Francisco (1998)

Moore, J.F.: The death of competition leadership & strategy in the age of business ecosystems. Harper Business, New York (1996)

Osterwalder, A., Pigneur, Y.: Business model generation: a handbook for visionaries, game changers, and challengers. Willey (2009)

Pisano, G.P., Verganti, R.: Which kind of collaboration is right for you? The new leaders in innovation will be those who figure out the best way to leverage a network of outsiders. Harvard Business Review 86(12), 78–86 (2008)

Rowe, F., Marciniak, R., Clergeau, C.: The contribution of information technology to call center productivity an organizational design analysis. Information Technology & People 24(4), 336–361 (2011)

Simon, H.A.: The architecture of complexity. Proceedings of the American philosophical society 106(6), 467–482 (1962)

Srikhanth, K., Puranam, P.: Integrating distributed work: comparing task design, communication, and tacit coordination mechanisms. Strategic Management Journal 32(8), 849–875 (2010)

Snow, C.C., Fjeldstad, O.D., Lettl, C., Miles, R.E.: Organizing continuous product development and commercialization: the collaborative community of firms model. Journal of Product Innovation Management 28(1), 3–16 (2011)

Williamson, P.J., De Meyer, A.: Ecosystem advantage: how to successfully harness the power of partners. California Management Review 55(1), 24–46 (2012)

Business Models, Symbionts and Business Ecosystem: A Case Study from E-commerce Industry in China

Wei Wei, Wuxiang Zhu, and Guiping Lin

Abstract. The essence of a business model defines a transaction structure that involves stakeholders. This article examines the transactional features that exist between stakeholders, such as relationships and form. We also introduce the new concept of "symbiont" and analyze the aggregation of the focal firm's and stakeholders' business models. By proposing this innovative concept, we seek to bridge the gap between the macro Business Ecosystem and micro Business Model, and as such expand our ideas about business models. The concept of "symbiont" in the business model that we have created provides a shared coordination system for different business models under the same business ecosystem, making direct comparison between them possible. Moreover, it allows us to analyze a focal firm's business model from a micro perspective, which may clarify how to precisely and scientifically restructure or reform the focal firm based on the anatomic picture of the company. To explain this theory, our research focuses on China's E-commerce industry.

Keywords: Business Model, Symbionts, Business Ecosystem, Stakeholders.

Wei Wei
Associate Professor of Management
Associate Dean, HSBC Business School, Peking University, Peking, China
e-mail: weiwei@phbs.pku.edu.cn

Wuxiang Zhu
Professor of Finance, Finance Department, School of Economics and Management, Tsinghua University, Peking, China

Guiping Lin
Ph.D. Candidate of Economics,
HSBC Business School, Peking University, Peking, China

P.-J. Benghozi, D. Krob, & F. Rowe (Eds.): *Digital Enterp. Des. & Manage. 2013*, AISC 205, pp. 161–182.
DOI: 10.1007/978-3-642-37317-6_14

1 Introduction

Traditional management research studies the organizational structure of enterprises, the relationships between various departments and staff, and the nature of specific business and management activities. The micro perspective focuses on staff while the macro perspective centers on the organization and industry. The corresponding evaluation criteria comprise various operational indicators that can be summarized into capabilities that in turn are able to fully utilize existing resources.

Research into business models studies the transaction structure, specifically the relationships and modes that bind the focal enterprise and stakeholders. The research perspective can be narrowed to focus on an enterprise's internal stakeholders, for example, internal logistics, information and payment platforms (with independent investments/profits), specific parties' interests, and rights allocation. These items can be broadened to cover the entire business ecosystem, including the external stakeholders that exist under the focal enterprise, such as suppliers, partners, customers, and competitors. We can also include the stakeholders of the focal enterprise's stakeholders, such as the suppliers and customers of its suppliers and customers, as well as its competitors' partners. The evaluation criteria represent the value and efficiency that is thus created. We can observe how value is created and transferred among stakeholders, and how much is acquired by the focal enterprise.

Expanding our perspective offers the following three benefits: (1) It defines the source of focal enterprise value and the value transfer path that runs through macro structures such as the business ecosystem; (2) It analyzes the focal enterprise from a micro perspective (internal and external stakeholders) to optimize the enterprise; (3) Unlike the previous comparison models of enterprises in the same sector or with the same industry background, the symbiont perspective creates a universal coordination system that can compare the varied business models operated by different enterprises.

The key to connecting the micro perspective – the business models of the focal enterprise - with the macro view involves creating a 'middle' research perspective. This is the focal point of our research: the symbiont prospective.

As an industry that has emerged rapidly over the past few years, e-business provides an ideal research subject for the following reasons:

The Internet and mobile Internet represent the future direction of business development. E-business will play a key role in this process;

Multiple typical business models have been created as e-business continues to evolve, providing an ideal background for the study.

2 Literature Review

Business models are highly important in both the business and investment sectors (IBM, 2008). However, related academic study remains sluggish due the lack of precise definitions (Zott, Amit & Massa, 2011). Some scholars define business models as the logic of value exchange and creation (Linder&Cantrell, 2000; Gordijn&Akkermans, 2001; Petrovic, Kittl&Teksten, 2001; Osterwalder, 2004), while others define them as the transaction relationship between enterprises and various stakeholders (Weill &Vitale, 2001; Amit&Zott, 2001). Recent research, Zott and Amit (2008) defines the transaction relationship as a system that comprises detailed activity systems implemented by the focal enterprise or its partners, the connection mode of these activities, and the stakeholders responsible for the operation and management systems. Wei & Zhu (2007) defined business models for the first time as transaction structures that involve stakeholders, and proposed in their latest research that business models mainly comprise transaction subjects, content, modes and pricing strategies (Wei, Zhu & Lin, 2012).

Wei, Zhu & Lin (2012) expanded the organizational, legal and corporate governance borders of the stakeholder's perspective to connect to previous research. Stakeholder theory mainly applies to corporate governance, and its primary purpose is to specify enterprises' social responsibilities. Stakeholder theory believes that enterprises should pay attention not only to related shareholders and creditors, but also to the customers, staff, communities and governments to whom they assume charitable, moral and legal responsibilities. (Freeman, 1984; Carroll, 1993)

Opinions are widely divided over the definition of stakeholders. Unlike the theory that applies to the corporate governance field, Wei, Zhu & Lin (2012) refined the definition of a stakeholder as a party that has independent interest demands and resources and is involved in transactions with the focal enterprise.

Defining the border of what constitutes stakeholders also varies among scholars, ranging from core stakeholders that directly work with enterprises to the entire business ecosystem (Moore, 1993). The business ecosystem concept comprises both the business activity and the ecosystem. It holds that the business ecosystem is very similar to the bio-ecosystem in terms of self-organization, self-adaptation, co-evolution and emergence (Peltoniemi and Vuori, 2004). Initially, the business ecosystem included "4P3S", i.e. people, place, product, process, structure, shareowner and society (Moore, 1996); in recent years, it has begun to incorporate business models (Tian, et al, 2008).

We believe that it is necessary to establish a suitable border for stakeholders that lie between the narrow definition of core stakeholders and the broader view that covers all the stakeholders in the entire business ecosystem. In this article, we define the stakeholder border point as the ‘symbiont border’.

Like the business ecosystem, the concept of symbiont originated from biology and is used to describe creatures of different kinds that co-exist, co-evolve or restrain each other (Ahmadjian, 1986). In recent years, it has become a philosophical concept and an independent discipline that focuses on symbiont units, modes, and environments. The theory is widely used in urban design (Kisho Kurokawa, 1987) and global competition (LEE, Seung-ryul, 2005). Unlike symbiont theory, the symbiont concept proposed in this article mainly focuses on the general transaction structure created by the focal enterprise and their stakeholders to co-exist and evolve, and is closer to the biological meaning.

It differs from the strategic alliance concept, the definition of which still varies widely among scholars. Some define it as the scope of the value chain that connects enterprises (Porter, 1997), while others view it as a transitional governance structure (Williamson, 1991). Overall, most agree that a strategic alliance has the following characteristics: complementary resources, shared risks and interests, long-term dynamic contracts and exclusivity (Yoshino and Rangan, 1995; Spekman et al, 1998; Borys and Jemison, 1989). Unlike a strategic alliance, the symbionts defined here comprises short- and long-term cooperation, resource-balancing and, above all, business models that include value creation and transfer structures and relationships, rather than a governance structure or exclusiveness.

3 Three Typical E-business Models in China

There are three typical e-business models in China: direct manufacturer sales (TCL), intermediary e-business (Dangdang) and third-party platform e-business (Alibaba).

3.1 Direct Manufacturer Sales

Some electric home appliance manufacturers, such as TCL, concurrently run online and offline stores.

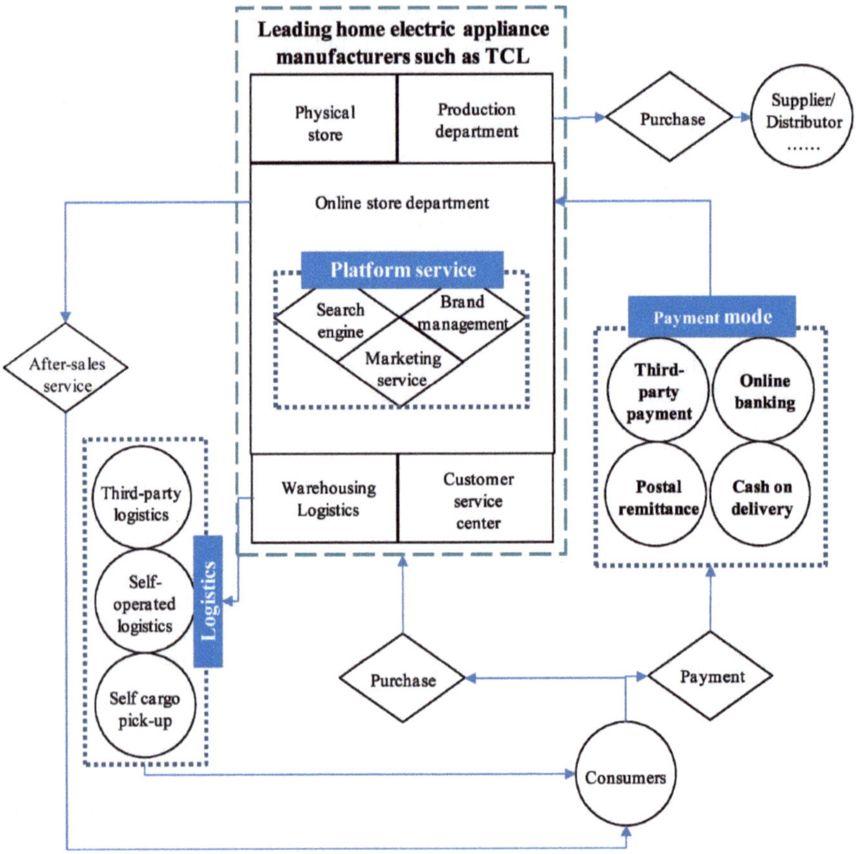

Fig. 1 Director manufacturer sales

These types of manufacturers provide search engines and brand management and marketing services for consumers through e-business channels, enabling consumers to select and pay for products using various payment modes, for example, third-party payment platforms, online banking, postal remittance and cash on delivery. They also provide various logistics modes such as third-party logistics, self-operated logistics, and self cargo pick-up, as well as various after-sales services.

These companies' e-business activities mainly involve managing offline stores, manufacturing, search engines, brand management, marketing, warehousing and logistics, and customer service centers on an e-business platform.

3.2 *Intermediary E-business*

Intermediary e-businesses such as Dangdang purchase products from suppliers and distributors, and provide e-business services including search engines, brand management and marketing services to enable consumers to select and pay for

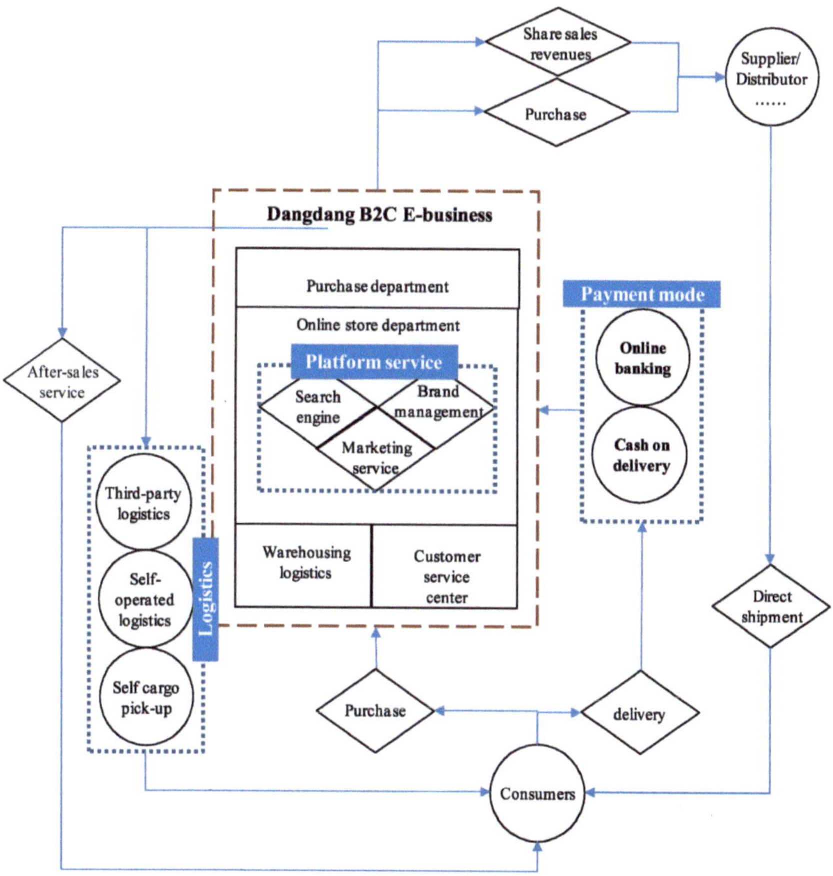

Fig. 2 Intermediary e-business system

products through online banking or cash on delivery. The products are delivered to consumers by third-party logistics, self-operated logistics or self cargo pick-up.

Dangdang has established a channel for suppliers to directly conduct transaction with consumers, much like the third-party e-business mode.

The e-business activities of intermediary e-business include purchasing and online store management (search engines, brand management and marketing), warehousing and customer service centers.

3.3 Third-Party Platform E-business

Third-party platforms such as Alibaba provide only the e-business channels, marketing and payment tools. Online stores can use these paid or free tools to display their products (usually purchased from suppliers or distributors). Consumers can pay for their products through Alipay, online banking or cash on delivery. Online stores usually use third-party logistics for customers and provide after-sales services.

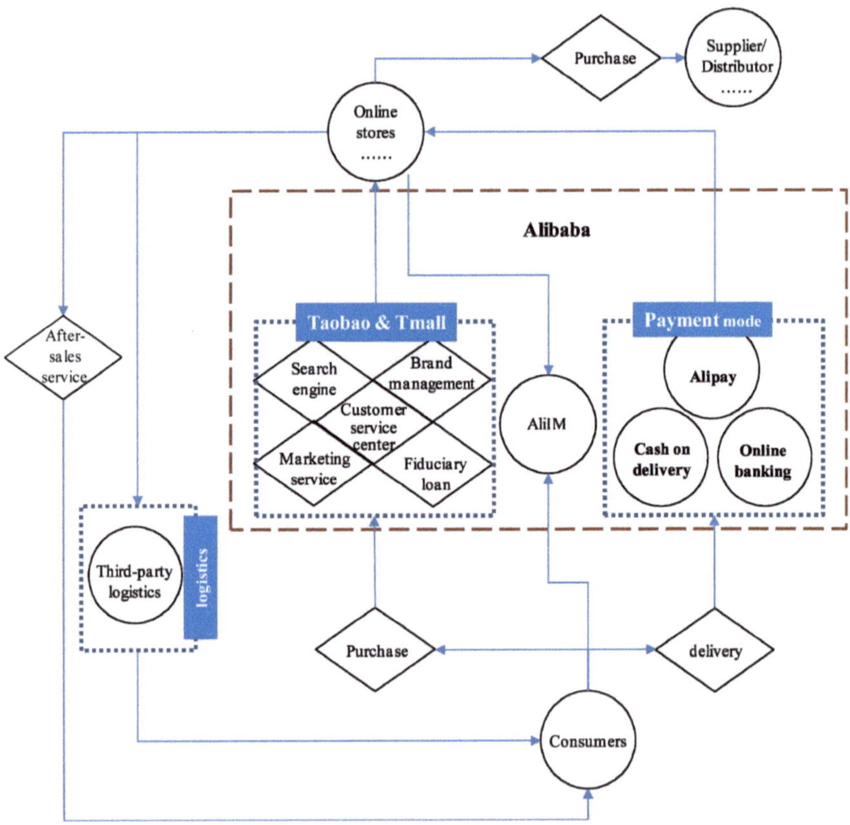

Fig. 3 Third-party platform e-business system

The business activities of third-party platforms involve online promotional channels (Taobao & Tmall, which includes e-business services such as search engines, brand management, marketing, customer services and credit); AliIM; and payment modes (including Alipay, cash on delivery and online banking).

The system figures for the above three e-business models reveal many common areas in terms of stakeholders, business, and management activities because all three belong to one symbiont.

4 Symbiont: Concept and Relationship with Business Models and the Ecosystem

The symbiont comprises the business models of the focal enterprise and part of the business models of its stakeholders.

For example, in the third-party e-business scenario mentioned above, the focal enterprise Alibaba (Taobao and Tmall) transacts with online stores and consumers, and the main activity system comprises network display channels, AliIM, and payment modes based on business model analysis.

Concept 1
Business models are the transaction structures that involve focal enterprises and their stakeholders

To expand the symbiont analysis area, stakeholders' business models must be analyzed. For example, online stores purchase products from suppliers and distributors and deliver them to consumers through third-party logistics companies. In addition to the focal enterprise and its stakeholders, a transaction structure also involves stakeholders' stakeholders, such as customers' customers, suppliers' suppliers and competitors. The symbionts also include internal stakeholders that have independent outputs, interest demands and rights distribution, such as logistics companies and information and payment platforms. The border of this definition is flexible. The overall transaction structure comprises the business model of the focal enterprise and part of the business models of the stakeholders.

Concept 2
The symbiont comprises the business models of the focal enterprise and part of the business models of its stakeholders

If we study the three e-business models using symbiont analysis, we discover that the three models belong to the same symbiont.

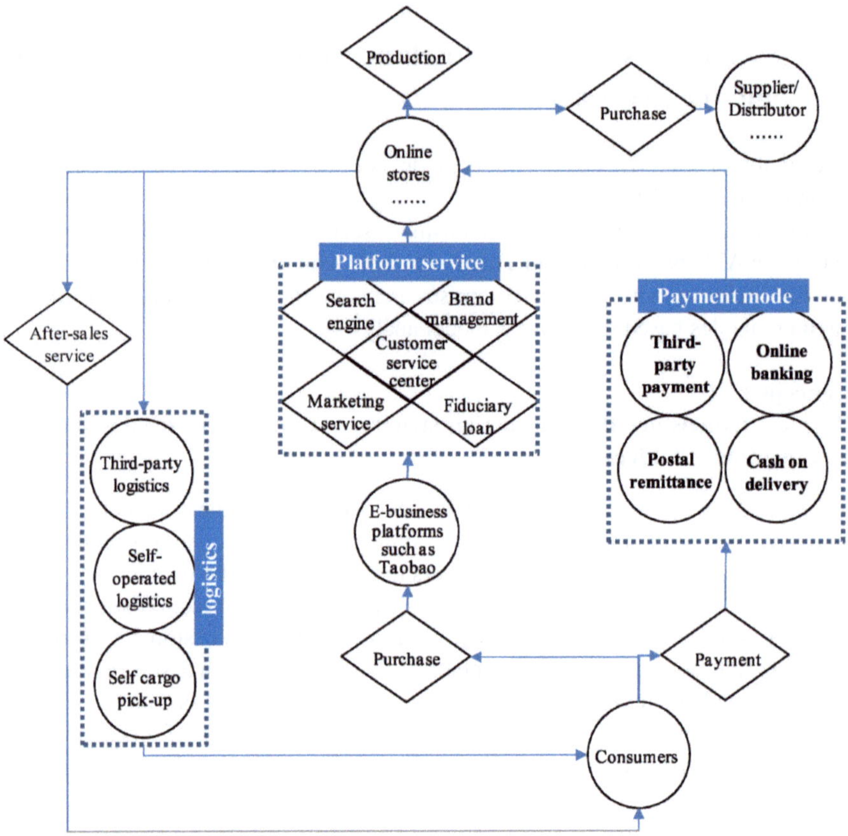

Fig. 4 E-business symbiont system

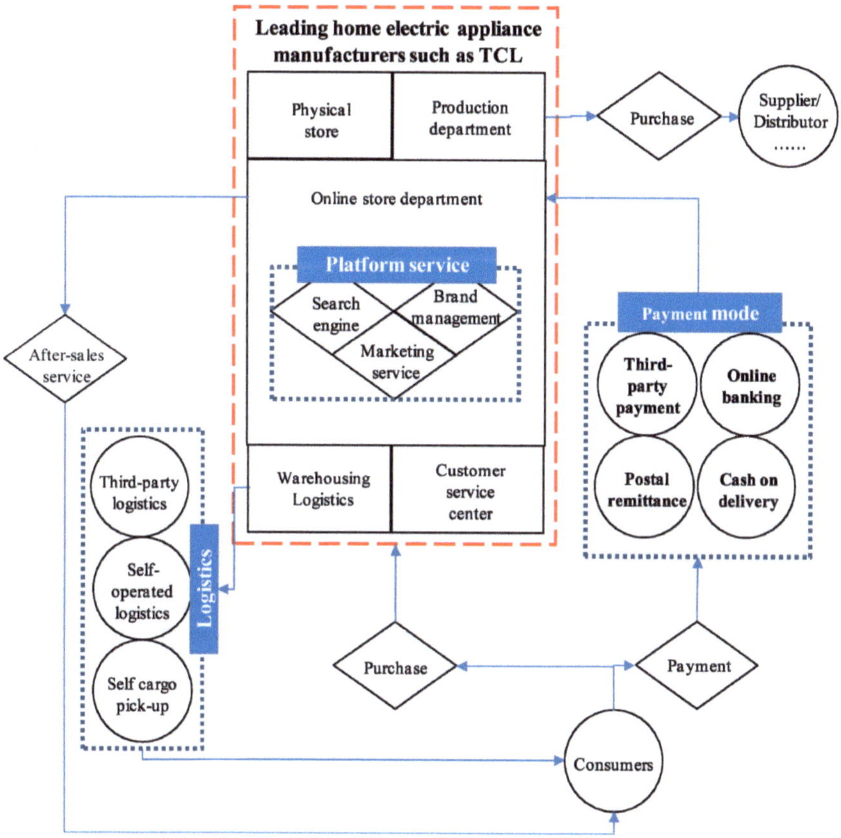

Fig. 5 E-business symbiont – direct manufacturer e-business system (the area enclosed by the red box is the focal enterprise)

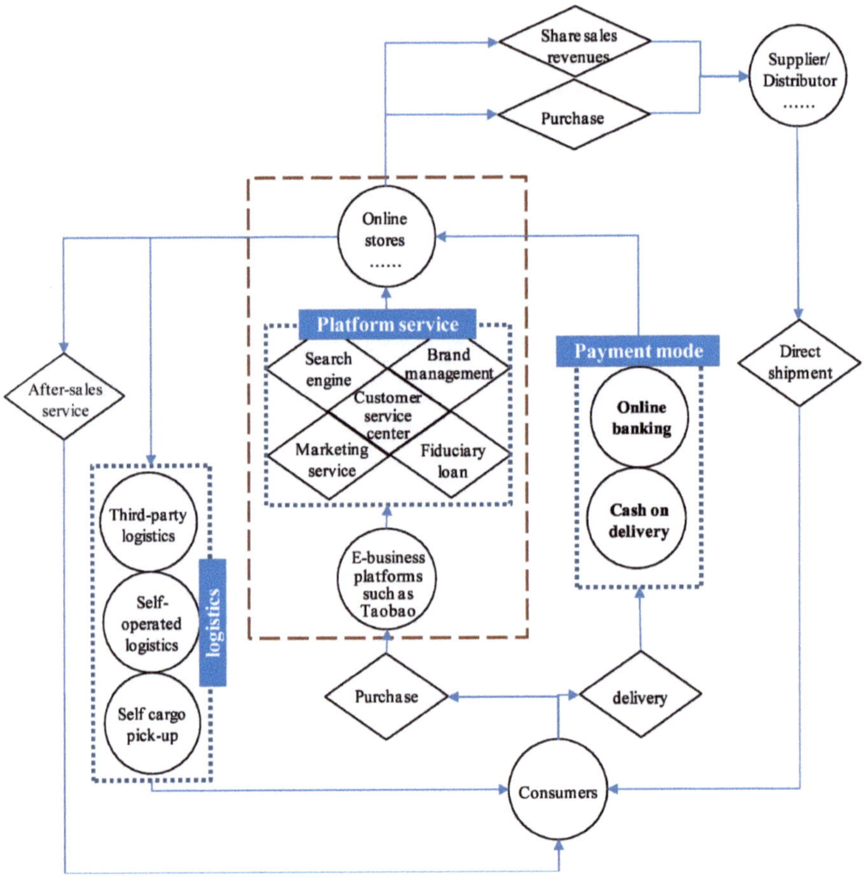

Fig. 6 E-business symbiont – intermediary e-business system (the area enclosed by the red box is the focal enterprise)

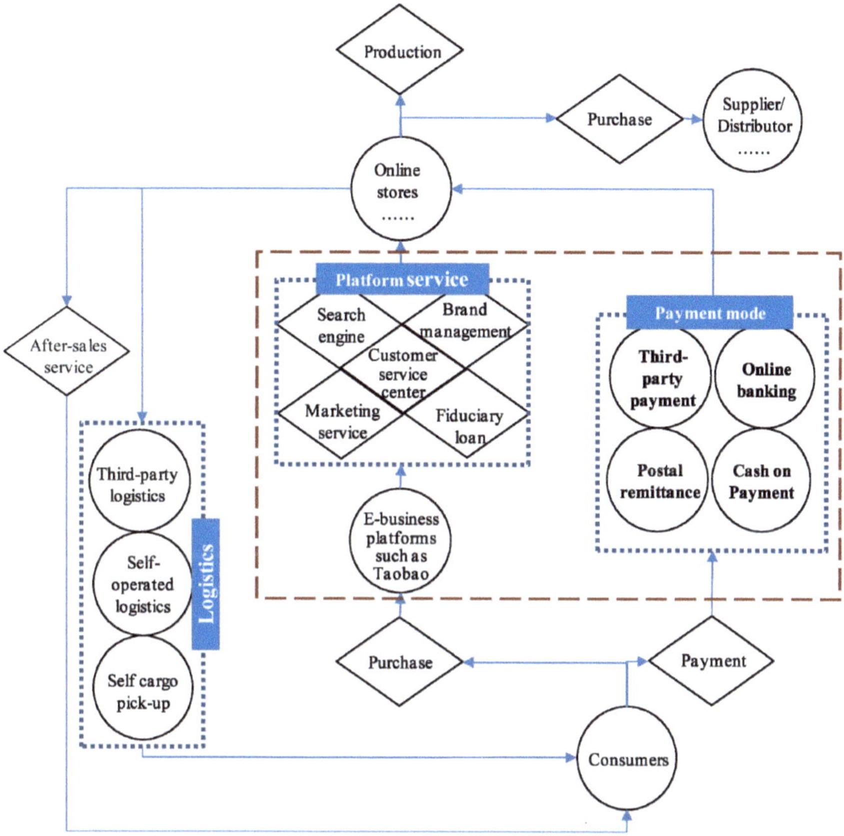

Fig. 7 E-business symbiont – third-party platform e-business system (the area enclosed by the red box is the focal enterprise)

When a symbiont involves the business models of specific stakeholders, only the business models that relate to the focal enterprise are analyzed; i.e. those that involve direct, indirect or potential transactions with the focal enterprise. Therefore, a symbiont only involves part of the stakeholders' business models.

Concept 3

The business ecosystem comprises the focal enterprise's symbiont, and the symbionts of its competitors (including similar products and substitutes), partners, and upstream and downstream stakeholders.

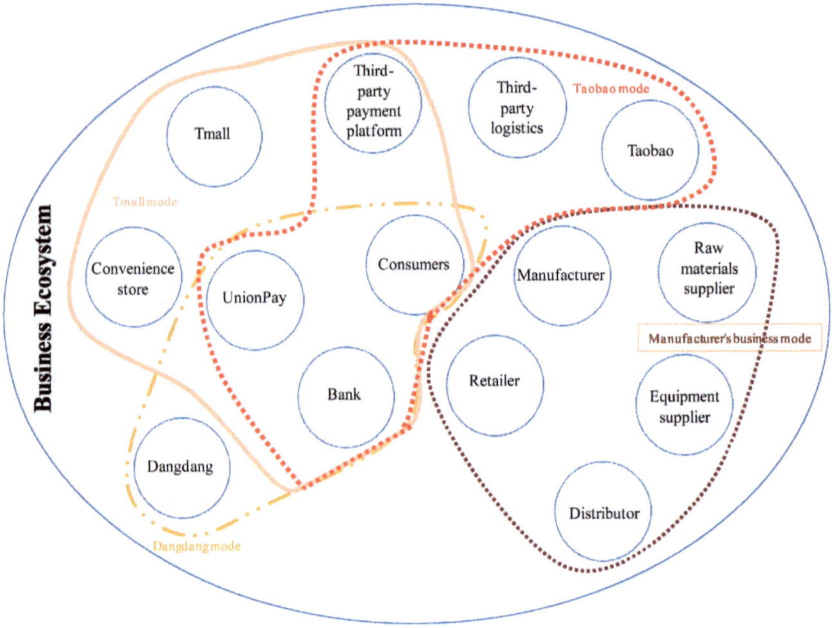

Fig. 8 Scope of stakeholders[1]: business models, symbiont and business ecosystem

The business ecosystem is another term that is often discussed with the focal enterprise's business model and symbiont. In essence, the business ecosystem comprises the symbiont, which in turn comprises the focal enterprise's business model.

The business model comprises the focal enterprise, stakeholders and their activity systems.

In addition to the focal enterprise's business model, the symbiont includes part of the business models of the stakeholder that have direct, indirect and potential transactions with the focal enterprise. For example, the business model of a third-party platform involves online stores and consumers, while the symbiont of the third-party platform includes third-party logistic companies, suppliers and distributors, and the business models of online stores. The business models of online stores also include the models of the banks and tax agencies that are linked to the third-party platform. Therefore, the symbiont fuses the third-party's business models and part of the business models of the online stores.

In addition to the focal enterprise's symbiont, the business ecosystem comprises symbionts of competitors that produce similar products, substitutes of their products, and upstream and downstream stakeholders. As shown in the above

[1] The activity system has a similar definition. To simplify the figure, the activity system is omitted.

figure, the business ecosystem comprises the symbionts of the three typical business models.

If the stakeholders in some symbionts are involved in one industry, their symbionts and those of their competitors form an industry ecosystem. The cross-industry symbionts formed based on such industry ecosystems form a business ecosystem.

5 Business Model Innovation from the Symbiont Perspective

1 Expanding from stakeholders to their stakeholders or further;

From the symbiont prospective, we can innovate a third-party platform e-business model to cover our customers' customers, suppliers' suppliers, and customers' partners, and then compare this model with the direct manufacturer and intermediary e-business models. This type of comprehensive macro perspective helps us with business model innovation and restructuring (see the following figure).

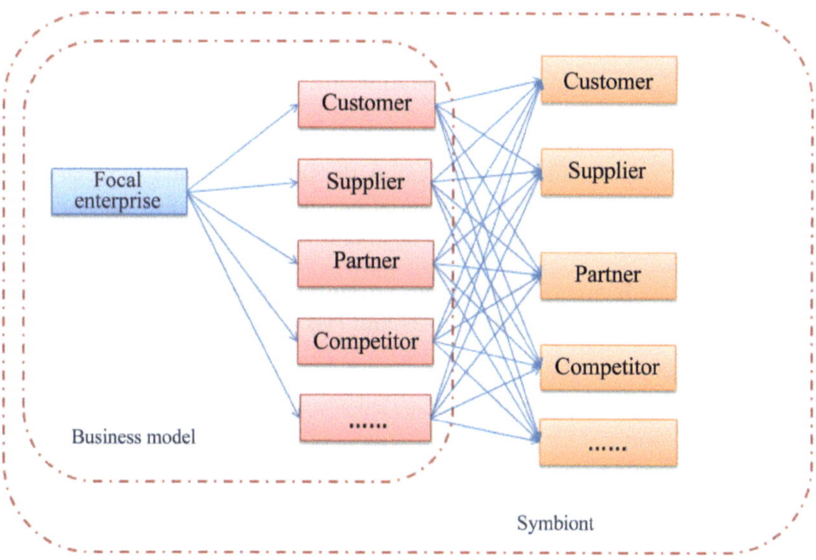

Fig. 9 Expansion of stakeholders (from a business model to the symbiont)

2 Expanding the existing value space to the entire business ecosystem

As shown the following figure, a symbiont creates transaction value for the stakeholders involved in the symbiont, and these stakeholders pay the corresponding transaction costs[2]. The margin between value and costs is called the value space. In addition to transaction costs, the focal enterprise and stakeholders pay monetary costs[3] such as raw material costs. The appreciated value is equal to the value space minus the monetary cost, or the surplus of the focal enterprise (enterprise value of the focal enterprise) plus the surplus of the stakeholders.

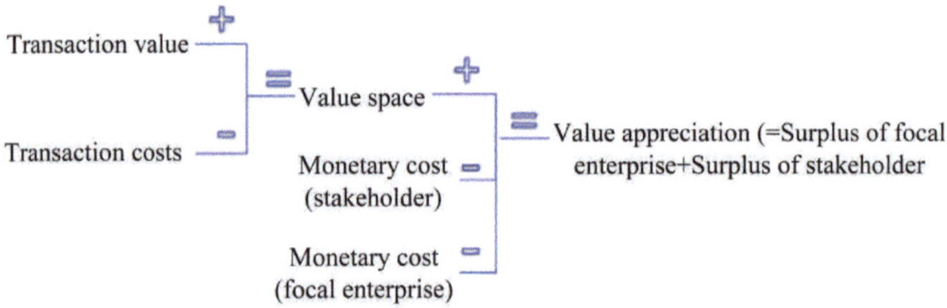

Fig. 10 Value space and appreciated value

From the perspective of an entire symbiont, the final sales value of e-business equals the total revenue from the platform and the logistics and payment services: i.e. the appreciated value shown in the following figure (e-business sales value = unit price x average sales volume of each online store x total number of online stores) minus the transaction and monetary costs incurred during production, transportation and sales. It also represents the maximum enterprise value that can be obtained.

3 Changing the business model from pursuing business scale and profits to integrating transaction value, costs and risks

To maximize value for an enterprise, e-businesses can increase transaction value by raising unit price, the average sales volume of each online store or the number of online stores; reducing transaction costs such as coordination, management and borrowing costs, and tax; or lowering monetary costs such as production and logistic costs. If a new business model raises the transaction value, both transaction and monetary cost rise in parallel. A new business model is effective only when the rise in transaction value exceeds the rise in transaction and monetary costs.

[2] Here we define transaction costs as costs caused during transactions.

[3] Here we define monetary cost as cost generated by the transaction object.

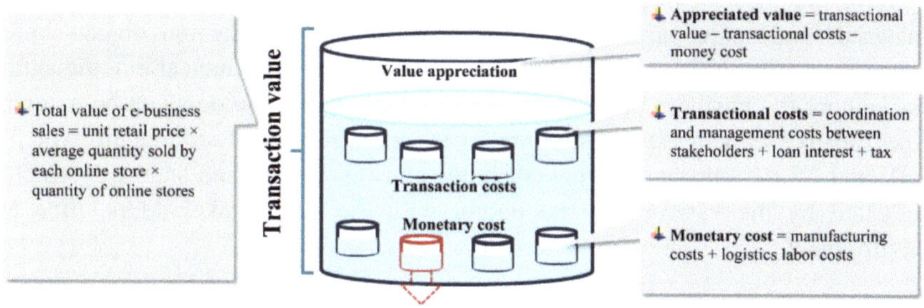

Fig. 11 Appreciated value of e-business eco-system

4 Expanding from existing stakeholders to new stakeholders

The symbiont concept provides a wider perspective for entrepreneurs to find resource weaknesses and design-in new stakeholders to compensate.

For example, Taobao has introduced online stores in third-party platform mode. Given the huge number of scattered stores, Taobao introduces agents to reduce business and transaction difficulties. The introduction of new stakeholders such as business agents substantially elevates Taobao's competitiveness.

Here we assume that the three e-business models utilize the same technologies, management methods and infrastructure. However, if any of these factors differ, both the symbiont and value change, even if the stakeholders and activity system are the same.

6 Division and Reorganization of Symbiont and Business Model Designing

Symbiont expands the prospective from the focal enterprise and its stakeholders to the entire business ecosystem, which spans the whole activity and industry value chains. The entire business ecosystem can be divided and reorganized in three aspects.

Any business and business activities involve investment, processing and output. Investment involves resources, processing reflects stakeholders' capabilities, ownership of output defines the stakeholders' roles (functions, powers and resource capabilities).

These elements, investment, activities[4] and output[5], can be divided.

[4] It is to divide an activity into several small activities, such as the three activities shown in the figure.

[5] It is to further divide the small activities into smaller ones.

Here, we must distinguish resources from capabilities. Resources include raw materials and spare parts to be invested, which are transaction objects before processing. Capabilities are stakeholders' properties that may affect the output throughout the business process. Such properties of stakeholders can be assessed. Specifically, we can assess such properties in two aspects: strengthen, which is reflected by the impacts of the stakeholder on the output, and stability, which is reflected by the variance in the output with the same stakeholders, time and environment.

Fig. 12 Three divisions

Divided resources, capabilities, businesses, management activities, and outputs (roles) can be reorganized to form new stakeholders that create a new business model.

The symbiont is an activity system that comprises stakeholders who operate different business models, reflecting their different relationships with the business system. Business systems and stakeholders provide different perspectives through which to analyze symbionts and business models.

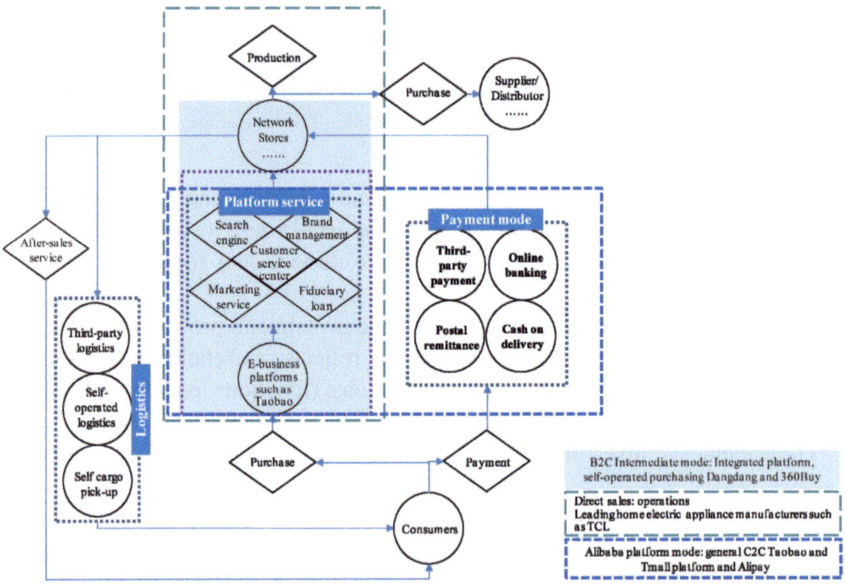

Fig. 13 Three business models for e-business (in the same symbiont)

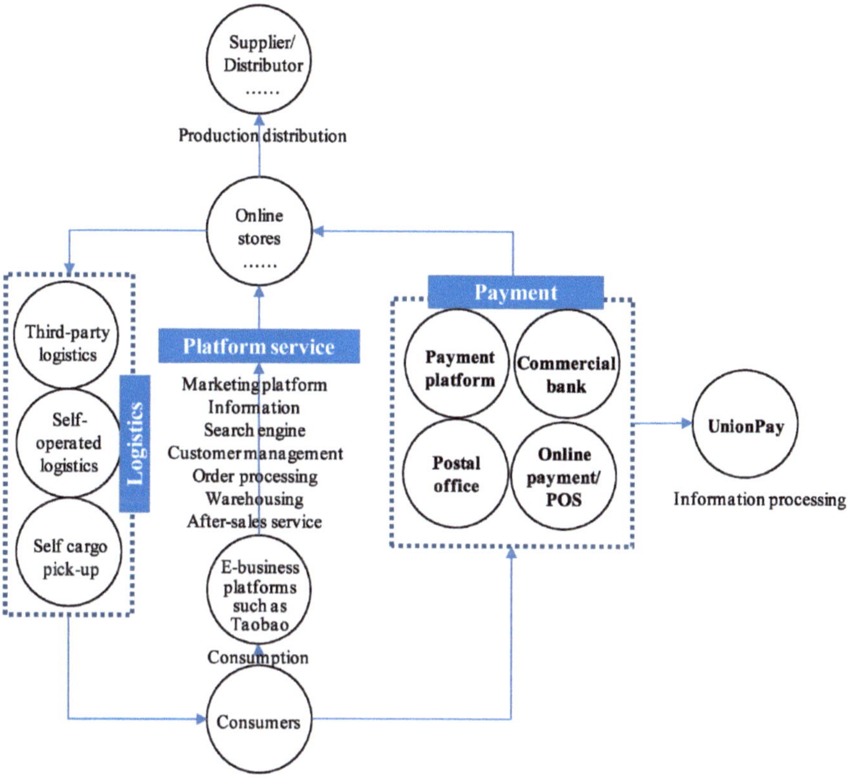

Fig. 14 Stakeholders and activities in the e-business symbiont

The relationships between stakeholders and activities are displayed in the following chart. The first row shows the divided activities and the rows below contain the corresponding stakeholders.

Division and reorganization of roles and resources are similar with the division of activities.

Division and reorganization can be assessed by three elements: transaction value, costs and risks. The division and reorganization of a business activity, role and resource capability will create specific transaction value, costs and risks. To elevate the business model's efficiency, the division and reorganization must realize at least one of the following targets: elevating the transaction value; lowering transaction costs; reducing transaction risks; or realizing a net increase in the transaction value.

The symbiont perspective is used to design business models in the following three scenarios

Analyzing the stakeholders and activities involved in a symbiont provides entrepreneurs with a full view of the business ecosystem. After deciding which activity to operate, an entrepreneur can design a business model to reflect its relationship with the activity.

Entrepreneurs can design strong business models that promote value appreciation in the symbiont in the following ways:

Elevating transaction value: increasing stakeholders (for example, iPhone's iTunes and App stores), expanding the business scale of similar stakeholders (chains and bilateral platforms), and increasing the demand of similar stakeholders (group purchases and complete solutions).

Table 1 Relationships between activities and stakeholders

Business links	Online stores	E-business platform	Independent e-business	Supplier/distributor	Third-party payment	Online banking	UnionPay	Physical store	Logistic company
Production				■					
Search engine	■	■	■					■	
Marketing service	■	■	■					■	■
Purchase				■				■	
Payment					■	■	■		
Logistics	■	■	■						■
After-sales service	■							■	
Communication between buyer and vendor	■		■					■	
Warehousing	■	■	■	■				■	■
Customer management	■	■	■			■	■	■	
Financial guarantee					■				
Credit and repayment							■		
Intermediary payment agent					■	■	■		

Lowering transaction costs: standardizing modules, integrating the backstage, merging similar transactions, and controlling governance structure and resource capabilities;

Reducing transaction costs: dividing and transferring risks through real object options and profit mode design;

More effective symbionts can be designed by analyzing, dividing and reorganizing the existing symbiont's transaction costs and risks.

7 Three Symbiont Theories

The three e-business models have the same stakeholders and activities. What can we learn by comparing the changes to symbiont efficiency under different business models? To answer this question, we have proposed the following three theories:

1 The efficiency of a non-dissipative symbiont is unaffected by different models

A non-dissipative symbiont must meet the following two conditions:

(1) Each transaction process incurs transaction costs that are transferred to one or multiple activities and stakeholders within the symbiont;

(2) Each transaction object includes multiple properties[6], each of which is fully priced and traded internally.

This transaction process meets the first condition because the value created by a non-dissipative symbiont is only transferred within the business system and between the stakeholders in that symbiont.

Different business models reflect the changes in relationships between business activities and stakeholders. The efficiency of a symbiont is assessed by transaction value, costs and risks.

In the ideal non-dissipative scenario in which the symbiont defines the extension of stakeholders (resources and capabilities) and business activities, business models only vary in terms of the relationships between business activities and stakeholders. A symbiont has no internal value dissipation such as transaction costs and unmatched capabilities and rights. Given the definition of the symbiont's border and the lack of external value exchange, the efficiency of the symbiont remains unaffected.

2 The efficiency of a dissipative symbiont is affected by different business models[7]

Dissipation in a symbiont occurs for two reasons:

Transaction costs can be divided in two ways: (1) They are transferred to one or more business activities and stakeholders in the symbiont; (2) They are transferred externally or dissipated, creating a non-zero result.

Some transaction object properties are either not fully priced or externalized during the transaction process, which impairs value.

If the symbiont's border is defined through dissipation when enterprises adopt different business models, the transaction costs and risks caused by transaction division and reorganization between stakeholders and the value dissipation in the symbiont will be different, which in turn alters the symbiont's efficiency. For example, assume that a third party provides services on a third-party e-business platform for an online store that holds fully priced operating capabilities and obtains all the profits from independent operations. This mode lowers value impairment and is more efficient than the intermediary e-business mode because the latter has to

---

[6] Such as quality, function, outer appearance, materials and experience.

[7] Note: If different business models possess the same dissipation features, the symbionts will have the same efficiency level; however, we assume that dissipation is different here because this scenario is unlikely in reality.

balance operational and e-business services. Therefore, the global scale of this type of third-party e-business platform is greater than other e-business models.

This culminates in our third theory:

3 An enterprise in a dissipative symbiont adopts the most efficient initial business model, resulting in efficiency gains that are clearly higher than the original or any alternative business model

The theory shows that changing the business model incurs costs, and that this must be considered by enterprises when selecting and designing effective business models.

Given the equal dissipation involved in establishing different business models, a symbiont's efficiency depends on the transactions within the business model and the dissipation caused by changing it. Changing a business model causes business activities and stakeholders to divide and reorganize, and triggers value dissipation, which can be neglected. Initially choosing the most efficient[8] business model creates higher efficiency than modifying or restructuring an original business model because doing so avoids value dissipation from division and reorganization.

Even if the existing business model is inefficient, it should not be changed if the costs of division or reorganization are excessive or if the resulting increase in value dissipation exceeds the subsequent transaction value. Moreover, if the total value dissipation of the symbiont exceeds its transaction value, the transaction structure between the focal enterprise and stakeholders will collapse unless external value is introduced into the symbiont.

Business models can be divided and reorganized. Specifically, division divides resources, activities (business and management activities) and outputs, while reorganization redefines the relationships between activities and stakeholders.

Modifying a business model must meet at least one of the following three targets: increasing transaction value, lowering transaction costs and risks, or elevating efficiency (determined by the transaction value, costs and risks).

8 Conclusion and Outlook

Research on business models expands the perspective of enterprises and their stakeholders. Although the business ecosystem provides a macro research perspective, it lacks a middle ground and a unified reference system for comparing and innovating business models. The theory of symbiont fills this gap.

Symbiont theory offers the following three benefits: (1) It defines the source of the focal enterprise's value and the value transfer path through macro structures such as the business ecosystem; (2) It analyzes the focal enterprise from a micro perspective (internal and external stakeholders) to enable optimization; (3) The middle-ground perspective of the symbiont establishes a unified reference system

[8] The scenario of establishing a business model through equal dissipation maximizes transaction value.

for enterprises that have adopted different business models within the same ecosystem, which in turn enables different business models to be compared.

Research on the symbiont will develop by establishing the following:

1 A mathematical symbiont model coupled with a theoretical deduction mechanism
2 Logic for business model comparison and a fiscal reference system that is underpinned by the symbiont concept
3 A symbiont perspective and a design path for business models
4 A method for boosting the efficiency of symbionts and the value of focal enterprises based on transaction value, costs and risks

(About the authors: Wei WEI, Peking University HSBC Business School; Wuxiang ZHU, Tsinghua University School of Economics and Management; Guiping LIN, Peking University HSBC Business School).

References

[1] Ahmadjian, V.: Symbiosis: an Introduction to Biological Association. University Press of New England, Englana (1986)
[2] Amit, R., Zott, C.: Value Creation in E-business. Strategic Management Journal 22(6/7), 493–520 (2001)
[3] Borys, B., Jemison, D.B.: Hybrid Arrangements as Strategic Alliances: Theoretical Issues in Organizational Combinstiond. Academy of Management Review 14, 234–249 (1989)
[4] Carroll, A.B.: Business and Society: Ethics and Stakeholder Management, 2nd edn. South-West, Cincinnati (1993)
[5] Freeman, R.E.: Strategic Management: A Stakeholder Approach. Pitman, Boston (1984)
[6] Gordijn, J., Akkermans, J., van Vliet, J.: Designing and Evaluating E-Business Models. IEEE Intelligent Systems 16(4), 11–17 (2001)
[7] IBM, The Enterprise of the Future: The global CEO study 2008.IBM Global Business Services (2008)
[8] Linder, J., Cantrell, S.: Changing Business Models: Surveying the Landscape. Accenture Institute for Strategic Change (2000)
[9] Moore, J.F.: Predators and prey: A new ecology of competition. Harvard Business Review 71(3), 75–86 (1993)
[10] Mooer, J.F.: The death of competition: Leadership and strategy in the age of business ecosystem, p. 76. John Wiley & Sons Ltd., Boston (1996)
[11] Osterwalder , The business model ontology-A proposition in a design science approach. UniversitédeLausanne (2004)
[12] Peltoniemi, M., Vuori, E.K.: Business ecosystem as the new approach to complex adaptive business environments. Paper presented at Frontier of E-business Research 2004, Tampere, Finland, pp. 1–34(2004)
[13] Petrovic, O., Kittl, C., Teksten, R.D.: Developing business models for ebusiness. Paper presented at the International Conference on Electronic Commerce, Vienna (2001)
[14] Spekman, R.E., Forbes, T.M., Isabella, L.A.: Alliance Management: A view from the Past and look to the Future. Journal of Management Studies 35(6) (1998)

[15] Tian, et al.: BEAM: A framework for business ecosystem analysis and modeling. IBM Systems Journal 47(1), 101–114 (2008)
[16] Weill, P., Vitale, M.R.: Place to space: migrating to e-business models. Harvard Business Press (2001)
[17] Williamson, O.E.: Strategizing, economizing, and economic organization. Strategic Management Journal 12(Winter Special Issue), 75–94 (1991)
[18] Yoshino, M., Rangan, U.S.: Strategic Alliances: An Entrepreneurial Approach to Globalization. Harvard Business School Press, Boston (1995)
[19] Zott, C., Amit, R.: The Fit Between Product Market Strategy And Business Model: Implications For Firm Performance. Strategic Management Journal (2008)
[20] Zott, C., Amit, R., Massa, L.: The Business Model: Recent Developments and Future Research. Journal of Management (2011)
[21] Kurokawa, K.: New Symbiosis. China Building Book Shop, Beijing (1987) (in Chinese)
[22] LEE Seung-ryul. The Era of Symbiosis – Northeast Asia Regional Development Roadmap. World Affairs Press, Beijing (2005) (in Chinese)
[23] Porter, M.E.: Competitive Strategy. Huaxia Publishing House, Beijing (1997) (in Chinese)
[24] Wei, W., Zhu, W.: Construction of Business Models. Shenzhen Science & Technology (March 2007) (in Chinese)
[25] Wei, W., Zhu, W., Lin, G.: Business Model Theory Based on the Stakeholder Transaction Structure. Management World (December 2012) (in Chinese)

Author Index

MIX
Papier aus verantwortungsvollen Quellen
Paper from responsible sources
FSC® C105338

If you have any concerns about our products,
you can contact us on
ProductSafety@springernature.com

In case Publisher is established outside the EU,
the EU authorized representative is:
Springer Nature Customer Service Center GmbH
Europaplatz 3, 69115 Heidelberg, Germany

Printed by Libri Plureos GmbH
in Hamburg, Germany